Impacts of Climate Change and Variability on Pastoralist Women in Sub-Saharan Africa

Editors

Munyae M. Mulinge

Melese Getu

Organisation for Social Science Research
in Eastern and Southern Africa (OSSREA)

FOUNTAIN PUBLISHERS
www.fountainpublishers.co.ug

Fountain Publishers
P.O. Box 488
Kampala
E-mail: sales@fountainpublishers.co.ug
publishing@fountainpublishers.co.ug
Website: www.fountainpublishers.co.ug

Organisation for Social Science Research in
Eastern and Southern Africa (OSSREA)
P.O. Box 31971
Addis Ababa, Ethiopia
E-mail: ossrea@ethionet.et
Website: http//www.ossrea.net

Distributed in Europe and Commonwealth countries outside Africa by:
African Books Collective Ltd,
P.O. Box 721,
Oxford OX1 9EN, UK.
Tel/Fax: +44(0) 1869 349110
E-mail: orders@africanbookscollective.com
Website: www.africanbookscollective.com

First published 2013

ISBN: 978-9970-25-236-7

Contents

Acknowledgements .... *v*

Acronyms .... *vi*

Chapter 1: Introduction 1
Munyae M. Mulinge

Chapter 2: Climate Change Impacts and Local Coping Strategies among Pastoral Women: Cases from the Southern Lowlands of Ethiopia 17
Aklilu Amsalu and Desalegn Wana

Chapter 3: Challenged Livelihoods as a Result of Water Scarcity among Maasai Women Pastoralists in Kajiado County, Kenya 45
Beth Njiru

Chapter 4: Gendered Impacts and Adaptation Mechanisms to Climate Change among Afar Pastoralists in North Eastern Ethiopia 83
Mulubrhan Balehegn and Kelemework Tafere

Chapter 5: Effects of Climate Change and Variability on Pastoralist Women's Accessibility to Social Services: Case of Fentalle Woreda, Central Rift Valley of Ethiopia 121
Getnet Feyissa

Chapter 6: Women Pastoralists and Climate Change Impacts in Kilosa District, Tanzania 155
Venosa A. Mushi

Chapter 7: Gender and Climate Change Adaptation among Agro-Pastoral Communities: Case Study of Chivi District in Southern Zimbabwe 187
Admire Mare

Chapter 8: Effects of Climate Change and Variability on Pastoral Communities: Coping and Adaptive Strategies for Women and Men in Uganda 223

Consolata Kabonesa and Fredrick Immanuel Kindi

Chapter 9: The Impacts of Climate Change and Variability on Pastoralist Women in Somali Region of Eastern Ethiopia..265

Bamlaku Tadesse

Chapter 10: Traditional Adaptation Mechanisms to Climate Change and Variability among Women Pastoralists in South-west Zimbabwe 297

Matsa Mark and Matsa Winniefridah

Acknowledgements

OSSREA's research and capacity-building programmes have been supported by SIDA, NORAD, DANIDA and the Royal Netherlands Government's Ministry of Foreign Affairs for the last three decades. OSSREA is indebted to all of these organisations for their continued support and the financial assistance that enabled this book to be published. We also thank the representatives of these organisations for their trust in our work.

We wish to thank all the authors of the book chapters, which are drawn from five sub-Saharan Africa countries. The authors worked hard not only to write the chapters but also to revise them many times to meet the stringent requirements of OSSREA. They have undergone through a long and a rigorous review process and the authors responded with diligence to the queries and comments of the editors during the editing process. The commitment and patience of all the authors is greatly appreciated.

Acronyms

AAP	African Adaptation Programme
ASALS	Arid and Semi-Arid Lands
BEAM	Basic Education Assistance Module
BNRCC	Building Nigeria's Response to Climate Change
CCAA	Climate Change Adaptation in Africa Programme
CDM	Clean Development Mechanism
CDO	Community Development Officer
CPR	Common Property Resources
CSA	Central Statistical Agency
DREA	Department of Rural Economy and Agriculture
FAO	Food and Agriculture Organisation
FDRE	Federal Democratic Republic of Ethiopia
FGD	Focus Group Discussion
FHH	Female-Headed Households
FTLRP	Fast-Track Land Reform Programme
GDP	Gross Domestic Product
GEF	Global Environment Facility
GIS	Geographical Information System
GoK	Government of Kenya
GOs	Governmental Organisations
GPS	Global Positioning System
HoAREC	Horn of Africa Regional Environment Centre
IASK	International Association for Scientific Knowledge
ICRISAT	International Crops Research Institute for Semi-Arid Tropics
IFAD	International Fund for Agricultural Development
IK	Indigenous Knowledge
IKS	Indigenous Knowledge Systems
IPCC	Inter-governmental Panel on Climate Change
IRIN	Integrated Regional Information Networks
IUCN	International Union for Conservation of Nature
LDCs	Least Developed Countries
LDCEG	Least Developed Countries Expert Group
MDGs	Millennium Development Goals
MHH	Male-Headed Households

MoE	Ministry of Education
NAADS	National Agricultural Advisory Services
NCAPD	National Coordinating Agency for Population and Development
NGOs	Non-Governmental Organisations
ODI	Overseas Development Institute
OECD	Organisation for Economic Co-operation and Development
OSSREA	Organisation for Social Science Research in Eastern and Souther Africa
PA	Peasant Association
PANNAR	Pannar Seed (Pvt) Ltd
PRA	Participatory Rural Appraisal
SEED CO	Seed co Limited
SSA	Sub-Saharan Africa
UK	United Kingdom
UN	United Nations
UNCCD	United Nations Convention to Combat Desertification
UNDP	United Nations Development Programme
UNEP	United Nations Environment Programme
UNESCO	United Nations Educational, Scientific and Cultural Organisation
UNFCCC	United Nations Framework Convention on Climate Change
UNHCR	United Nations High Commissioner for Refugees
UNICEF	United Nations Children's Fund
UN-OHRLIS	United Nations Office of the High Representative for the Least Developed Countries, Landlocked Developing Countries and Small Island Developing States
URT	United Republic of Tanzania
USAID	United States Agency for International Development
WEDO	Women's Environment and Development Organisation
WMO	World Meteorological Organisation
WWF	World-Wide Fund for Nature
ZNGP	Zimbabwe National Gender Policy

Chapter 1

Introduction

Munyae M. Mulinge

Climate Change: The Challenge of the 21st Century

The term climate change is used to denote any significant but extended change in the measures of climate. The changes could be due to natural variability or as a result of human activities, such as the burning of fossil fuels to produce energy, deforestation, industrial processes, and some agricultural practices (IPCC TAR 2001a; Levina and Tirpak 2006; EPA n.d). Such activities release large amounts of carbon dioxide and other greenhouse gases into the atmosphere that hang like a blanket around the earth, thus trapping energy in the atmosphere and causing it to warm up (EPA n.d). This results increasingly in climate variability, which is characterised by extreme seasonal, annual, temporal and non-spatial variability in temperature, vagaries of precipitation (rainfall patterns and amounts) and/or wind patterns occurring over a prolonged period of time (*Ibid.*).

Based on the 4th Inter-governmental Panel on Climate Change (IPCC) assessment report (IPCC 2007), the global mean surface temperature has increased by about 0.07°C for every decade in the past 100 years, with the increase becoming more rapid at about 0.18°C per decade in the last 25 years. The last decade (2001 – 2010) has been the warmest on record; with the average temperatures reaching 0.46°C, above the 1961 – 1990 mean, and 0.21°C warmer than the 1991 – 2000 period (AMCEN 2011). For Africa in particular, the 4th IPCC Assessment Report (IPCC 2007) indicates that the continent is warming up faster, all year-round, than the global average; a trend that is likely to continue. By the year 2100, it is predicted that temperature changes will fall into ranges of about 1.4°C to nearly 5.8°C increase in mean surface temperature compared to 1990, and the mean sea level will rise between 10cm to 90 cm (AMCEN 2011). The interior of semi-arid margins of the Sahara and central southern Africa will be the most affected by such warming (AMCEN 2011).

Changes in temperatures have been accompanied by changes in weather and climate that have resulted in a myriad of both direct and indirect as well as actual and potential problems and challenges. These are being manifested in increased frequency of extreme weather events, variability in precipitation, increasing severity of floods, more intense and prolonged droughts, accelerating and expanding desertification, a rise in sea temperatures, severe heat and cold waves, rising sea levels, mounting acidity of oceans, shrinking sea ice, glaciers, and snow cover, displacement of huge populations, changes in cultivation seasons, thus declining agricultural productivity and food security, as well as severe water shortages, and increases in diseases such as malaria, dengue fever and cholera, among others (Warren 2006; UNFCCC 2007; IPCC 2007; OECD 2008; Ziervogel and Ericksen 2010; The World Bank 2008b; EPA n.d.). The intensification of these and other changes will no doubt present challenges to human societies and their environments. In particular, they will affect water supplies, agriculture, power and transportation systems, the natural environment, and the health, welfare, and safety of humans (EPA n.d.).

To tackle the phenomenon of climate change effectively, human societies have put in place a combination of *mitigation* and *adaptation* mechanisms and strategies. Whereas *mitigation* aims at avoiding or lessening the impacts of the unmanageable, the goal of *adaptation* is to manage the unavoidable (The World Bank 2008b).

Mitigation strategies entail taking deliberate steps to counteract the risks from climate change by making choices that reduce greenhouse gas pollution. While there exists no consensus as to what specific steps and actions should be taken or mooted to combat climate change, world nations are unanimous that something must be done. This is evinced by the various global and regional conferences and protocols and/or treaties committing signatories to the fight against climate change. Leading the way in this direction was the United Nations Conference on Environment and Development (UNCED), held in Rio de Janeiro in 1992, which provided the vision and framework for the multi-lateral cooperation needed to achieve a sustainable future under a changing climate (see AMCEN 2011). Subsequent UN conferences on climate change have culminated in the Kyoto Protocol[1] to the United Nations Framework Convention on Climate Change (see 'Endnotes'), the Copenhagen Accord signed in Copenhagen, Denmark in 2009 (see UNFCCC 2009), the Cancun Agreements reached in December 2011 in Cancun, Mexico, at the 2010 United Nations Climate Change Conference (see UNFCCC 2010), and the Durban, South Africa, Agreement of 2011 (see UNFCCC 2012). Jointly, these represent the key steps forward made by the international community to address the long term challenges of climate change collectively and comprehensively by reducing greenhouse gas emissions and helping world nations to protect themselves from the impacts of climate change.

Other initiatives for the *mitigation* of climate change effects include investments in clean energy in the developing world to help their transition to a lower carbon development path, the adoption of less greenhouse gas-intensive technologies, searching for efficient transportation technologies so as to reduce carbon emissions, and adopting better forestry and agricultural techniques to improve livelihoods, reduce soil erosion, and protect bio-diversity (The World Bank 2008a; 2008b).

On the other hand, *Adaptation* refers to "changes in processes, practices, and structures to moderate potential damages or to benefit from opportunities associated with climate change" (AMCEN 2011, 52). It basically has to do with the adjustment in natural or human (eco)systems as a means for ameliorating the actual or anticipated adverse effects associated with climate change by moderating harm or exploiting beneficial opportunities (IPCC TAR 2001b). *Adaptation* entails grappling with the changes that are already underway by evolving coping mechanisms through taking practical steps to cushion oneself from the likely disruption and damage resulting from the effects of climate change (Levina and Tirpak 2006; IPCC 2007). As such, it aims at reducing vulnerability to climatic change and vulnerability of communities, regions, and nations to climate variability, and promoting sustainable development (IPCC 2001). *Adaptation* mechanisms may be structural, technological and behavioural changes. These could be anticipatory or reactive, private or public, autonomous or planned (IPCC 2007; AMCEN

2011). From a temporal perspective, *adaptation* interventions can be short or long term, localised or widespread (IPCC 2007). According to The World Bank (2008a), *adaptation* calls for more resilient infrastructure, broader disaster relief and preparedness measures, and new agricultural technologies and practices to counter the increased climate risks.

With regard to Africa, historically the debate over climate change has focused primarily on *adaptation* as opposed to *mitigation*, with practitioners focusing mainly on attracting finance to build the continent's adaptive capacity (AMCEN 2011). Nevertheless, some *mitigation* measures anchored on Article 3 of the United Nations Framework Convention on Climate Change, that is, the principle of equity and of "common but differentiated responsibilities and respective capabilities" (UNFCCC 2006) have been initiated. These have included the push for investments in clean energy, the adoption of less greenhouse gas-intensive technologies, the search for efficient transportation technologies characterised by reduced carbon emissions, and better forestry and agricultural techniques for the reduction of soil erosion and protection of the bio-diversity (The World Bank 2008a and 2008b; AMCEN 2011). Turning to conferences and declarations, the African Ministerial Conference on the Environment, held from 7th to 12th June 2008 in Johannesburg, South Africa, can be cited as a landmark *mitigation* effort on the continent. The conference deliberated on the continent's participation in the Clean Development Mechanism (CDM) established by the Kyoto Protocol. It called for the improvement of the CDM to guarantee geographical equity in the distribution of projects contributing to sustainable development efforts on the continent (AMCEN 2011). Its most principal outcome was the AMCEN Nairobi Declaration of 2009. As well, mention should also be made of the 3rd Africa Carbon Forum that was held in Morocco in 2011 with the motto "Marrakech Plus 10" (AMCEN 2011), among others. This event, which aimed to increase the number of CDM projects on the continent, was to become the region's premier carbon finance trade fair and conference as well as knowledge-sharing platform.

The Differential Effects of Climate Change

Despite increasing *mitigation* and *adaptation* initiatives across the globe in general and Africa in particular, the phenomenon of climate change and variability remains one of the major threats to economic growth and development the world over. Whilst it affects people of all colour or races across the world, its impacts are distributed disproportionately as manifested not only among different regions, but also in terms of level of economic development, ecosystems, age and gender (IPCC 2007; UNFCCC 2007; Kraub 2011). As such, certain regions and groups are more vulnerable to climate shocks and likely to experience more severe consequences than others. For instance, the effects of climate change are expected to hit developing countries the hardest (The World Bank 2008b). In Africa, climate change, variability, and associated growing disaster risks present an additional burden to sustainable development by threatening and impeding the attainment of the Millennium Development Goals (AMCEN 2011). Here, escalating temperatures, changing rainfall patterns, rising sea levels, and more frequent weather-related disasters pose risks for health, water supply and sanitation, agriculture, forestry, fisheries, food supply, energy, transport, industry, mining, construction, trade, tourism, environmental protection, and disaster management (The World Bank 2008b). This will in turn undermine any gains made in the fight against poverty, hunger and disease, thereby endangering the lives and livelihoods of billions of people (*Ibid.*).

> According to studies (Hesse and Cotula 2006; Oesterle 2008; Oxfam International 2008; Djoudi and Brockhaus 2011), in Africa, dry land ecosystems as well as pastoral communities inhabiting them are the most vulnerable to the impacts of climate change and variability. Although the pastoral way of life is dependent on the rearing of animals (cattle, camels, donkeys, sheep and goats), such areas are characterised by moisture stress, unreliable rainfall and fragile landscapes that affect availability of grass and pasture. The magnitude of this problem is best understood when one considers the fact that an estimated 50 million pastoralists live in sub-Saharan Africa inhabiting arid and semi-arid regions. The literature indicates that livelihoods of pastoral communities are dependent on climate-sensitive resources such as water and pasture. Pastoralists in Africa are also vulnerable people

> who often suffer repeated, multiple and mutually reinforcing shocks that affect their families, their settlements and their livelihoods due to famine, drought, floods, and other climate change-induced disasters. In these communities, women are the main actors in agricultural production and water collection – activities that are susceptible to risks of climate change and variability. Women are not only playing key roles in managing the natural environment but are also heavily affected by the outcomes of climate change and variability.

Concerning the impacts of climate change and variability vis-à-vis gender, various studies (Oxfam International 2005; Neumayer and Plümper 2007) document the nexus between the two. A London School of Economics study of disasters (cited in Neumayer and Plümper 2007) conducted in 141 countries shows that gender differences in deaths from natural disasters are directly linked to women's economic and social rights. Similarly, a report on the 2004 Asia Tsunami by Oxfam International (2005) raised alarms about the gender dimension of the problem as the majority of those killed and those least able to recover were found to be women. Some reports indicate that women provide up to 90 per cent of rural poor people's food and produce 60 - 80 per cent of the food in most developing countries. Nevertheless, they are insufficiently represented in decision-making processes on community-based climate change adaptation and mitigation strategies. The same reports also indicate that women's relative lack of education and access to information on projected climate change impacts, such as rainfall patterns, effective natural resource management strategies, such as efficient water use, crop diversification or rotational grazing, limit their ability to cope with or adapt to the effects of climate variability and change (*Ibid.*).

Among pastoral communities, the limited available evidence supports the view that women, relative to their male counterparts, suffer more from climate change-related problems (Denton 2002; IPCC 2007; UNFCCC 2007; Aklilu and Alebachew 2009; Anderson, Hesse and Nassef 2009). This can be understood in terms of the differential roles played by the sexes among pastoral communities. To illustrate this, women care for children and weak livestock, meaning that they are left behind whenever severe drought forces men to migrate to distant places in search of water and pasture. In addition, women are responsible for the procurement of food for the family and hence, highly dependent on the natural environment (Dankelman 2002). Therefore, any degradation in the environment would adversely affect them more than men (Dankelman 2002; UNFCCC 2007; Anderson, Hesse and Nassef 2009). Put another way, frequent droughts and/or heavy precipitation that may reduce the availability of food for the family will ultimately undermine women's ability to provide for their families or increase their burden more.

That men and women are affected differently by climate change suggests that the sexes also differ in terms of the adaptation mechanisms they employ. As Cannon (2002) testifies, men and women experience different vulnerabilities and play different roles in coping with natural disasters. In particular, women's informal rights to resources (especially to land and water), lack of finance, and lack of knowledge (awareness/understanding) due to their low levels of education, have a bearing on their adaptation and coping strategies (Mogotsi, Nyangito, and Nyariki 2011). For example, women are thought to rely more on traditional coping strategies. In this regard, they naturally travel long distances to gather wild fruits to provide for their families as well as grass and tree fodder to feed cattle and small goats and sheep (Conroy 2002).

Despite the existence of gender-based differences in the effects of climate change and in adaptation and coping strategies, studies on the link between climate change and variability in sub-Saharan Africa, in general, and southern and eastern Africa, in particular, are limited. In particular, there have not been any country-specific in-depth studies or comparative studies that assess the gender dimensions of climate change and variability, especially on pastoral communities. As such, no proper examination of the situation has been made on the important insights and knowledge that pastoralist women developed on climate change adaptation through their experiences in their community and households. Again, except for case studies from a few countries (Botswana, Mozambique, Namibia and South Africa), the positive contributions of women in relation to climate change adaptation are not yet well recognised, studied and documented. By the same token, women's intimate knowledge about resource management and adaptation skills has not been well researched in eastern and southern Africa regions.

Driven by the recognition and appreciation of the need to comprehensively understand the phenomenon of climate change and variability, more so its gender dimensions among pastoral societies, the Organisation for Social Science Research in Eastern and Southern Africa (OSSREA), a non-governmental and non-profit research network that was founded in 1980 and headquartered in Addis Ababa, Ethiopia, initiated a call for abstracts with the objective to fill this knowledge gap. In doing so, OSSREA was guided by one of its umbrella objectives (that is, to encourage and promote interest in research in the social sciences in Eastern and Southern Africa) and by the realisation that climate change was now a major factor in Africa's socio-economic development, more so among pastoral societies. The overall purpose of the call for abstracts was to generate quality research papers from academics in Eastern and Southern Africa that would form the contents of an edited book offering a comprehensive understanding of the gender dimensions of climate change and variability among pastoral societies in the two regions.

The call for abstracts required the researchers to design and execute empirical studies focusing on any of the following subjects:

- Gender-related risks, vulnerabilities and opportunities associated with climate change and variability on pastoralist communities;
- Gender-differentiated impacts of climate change and variability on pastoral communities;
- Level of awareness of pastoralist communities on the occurrence, severity, dynamics and impacts of climate change and variability on pastoral women;
- Pastoralist women's historical, traditional and new coping and adaptation mechanisms to climate change and variability and their dynamics;
- Success stories and challenges of policies, programmes and projects that are designed by national governments to address the negative impacts of climate change and variability on pastoral communities especially women; and

- The role of pastoral women in planning and decision-making on issues of climate change and variability at community and national levels.

In all, fourteen researchers responded to the call for papers. These were subjected to a rigorous review process designed to produce a rank ordering of all papers submitted based on their quality. The review keyed on the following aspects of the papers submitted:

- *Introduction/Background:* Emphasis was on clarity and relevance of the context of the research, the delineation of a clear research problem, the author's ability to situate the research problem in the light of the existing state of knowledge, and the specification of clear objectives for the study.
- *Literature Review:* Stress was placed on the depth, relevance, and currency and thematic organisation of the literature presented; its ability to provide clear support to key theoretical and methodological issues or questions being investigated; and the author's ability to identify areas of contestation and show the gap(s), articulate the review clearly and to demonstrate an in-depth grasp of the empirical and theoretical knowledge surrounding the subject of the study.
- *Research Design and Methodology:* The review focused on the suitability of the methodology chosen, justification of data collection methods and procedures, reliability and accuracy of data sources, and the extent of the researcher's grasp of research methods (techniques) and the research process in general.
- *Findings and Discussion:* This looked at how clearly and sufficiently the results have been profiled, the extent to which the results are systematically presented and anchored on the objectives of the study, the presentation of tables and figures, and the adequacy of interrogation of core findings (including the extent of integration of existing knowledge and how well interpretation is anchored on the objectives of the research or its study questions.
- *Conclusions and Recommendations:* Emphasis was placed on whether conclusions had been drawn logically from results and discussions; whether they reflected on the objectives, theory and conceptual framework of the study; the extent to which

conclusions and recommendations captured the focus of the study as well as the statement of the problem and the study objectives.

- *Referencing:* The review focused on the clarity and consistency of referencing, the completeness of the reference list, and the reliance on a conventionally approved style (for example, Harvard or APA).
- *Overall Quality, Presentation and Language:* Here the focus was on the structuring and paragraphing of the paper; the extent to which content was articulated clearly and in a systematic and coherent fashion; and the originality of the paper.

Out of the 14 papers submitted for consideration, eleven were judged to have the potential for publication during the first phase of the review process. The authors of these papers were then requested to revise them, guided by the comments of the peer reviewers, before re-submitting them for a second round of review. After the second round of peer review, it was concluded that nine (9) of the eleven papers considered were of publishable quality. It is thus these nine papers that formed the core of this book. However, as a measure of utmost quality assurance, the selected papers were subjected to further scrutiny by a content editor before publication.

It is hoped that the knowledge generated through the papers of this book is significant in two ways.

Firstly, the research papers contribute information that fills an important knowledge gap in our understanding of the consequences of and adaptation to climate change and variability among pastoral communities in Africa, in general, and among pastoralist women, in particular. To reiterate, the call for the abstracts leading to this book project was anchored on the realisation that this subject matter had not been properly and adequately interrogated. In particular, save for case studies conducted in a few countries such as Botswana, Mozambique, Namibia and South Africa, women's positive contributions to climate change adaptation remain unrecognised, unstudied and undocumented. Similarly, their intimate knowledge about resource management and adaptation skills remain under-researched.

Secondly, by expounding on the gendered aspects of adaptations to climate change and variability, the papers presented in this book contribute invaluable knowledge that can guide policy makers to develop adaption and intervention strategies to combat the negative effects of climate change. The knowledge will also guide other stakeholders, including the grassroots populations that are directly impacted by climate change and variability, as they struggle to cope and adapt to the phenomenon of climate change and variability.

This book is structured into ten chapters:

The present chapter (Chapter One) is an overview of the situation in the context of Africa.

The second chapter investigates the impacts of climate change on and the local coping strategies utilised by women pastoralists in Borena and South Omo zones in the southern lowlands of Ethiopia. The results reinforce the view that existing socio-cultural situations of the pastoral community have put women in a disadvantaged position primarily because of the asymmetrical power relationship between men and women over control of and access to resources, as well as decisions related to the utilisation of household resources.

Chapter Three keys on the role of water scarcity in the changing livelihoods of Maasai women pastoralists of Oldonyo-orok sub-location of Kajiado County in Kenya. In particular, it assesses how water scarcity has changed livelihood systems of the Maasai women pastoralists and their coping strategies, along with the effects the changed livelihood systems have had on resource use.

Chapter Four, "Gendered Impacts and Adaptation Mechanisms to Climate Change among Afar Pastoralists in North Eastern Ethiopia" utilises a combination of different research techniques to assess the differences in knowledge, perceptions and attitudes regarding climate change. It also looks into the impacts of climate change in the context of differential degree and level of vulnerability, and adaptive capacity existing between Afar men and women. The results indicate how gender inequality, an inherent customary tradition (*Adda*) among the Afar, contributes to the increase in the vulnerability of women to the ills of drought. The study also reveals that activities, such as labour migration, petty trading and individual wealth accumulation, hitherto alien to the Afar culture, have emerged due to climate change.

Chapter Five, which focuses on Fentalle *Woreda*, Central Rift Valley of Ethiopia, examines the effects of climate change and variability on pastoralist women's access to social services. The author argues that women in pastoral communities are more vulnerable and prime victims of climate change – a situation even more aggravated by multiple variables and social, economic, and geographical factors.

"Women Pastoralists and Climate Change Impacts in Kilosa District, Tanzania", in Chapter Six, discusses how climate change, especially drought and floods, affects women pastoralists in the district, and the coping and adaptation strategies they utilise to deal with the situation. The results show that, despite the fact that both men and women pastoralists are negatively affected by climate change and variability, women still stand out as the more vulnerable group and hence the more affected. This is mostly due to inequalities between men and women in terms of household gender division of labour, ownership of resources and decision-making.

In Chapter Seven, "Gender and Climate Change Adaptation among Agro-Pastoral Communities: Case Study of Chivi District in Southern Zimbabwe", the author discusses existing perceptions of communities on the occurrence and impact of climate change and variability, its gender differentiated impacts, as well as the traditional and new coping and adaptation mechanisms adopted by the agro-pastoral communities in the district. Arguing that the gendered impacts of climate change are relative and contingent upon factors such as household dynamics and the wider socio-cultural and economic circumstances in which the agro-pastoralists operate, the author recommends the implementation of five main strategies that would enable agro-pastoral communities to mitigate the impact of climate change in the area.

The eighth chapter, "Effects of Climate Change and Variability on Pastoral Communities: Coping and Adaptive Strategies for Women and Men in Uganda", interrogates the effects of climate change and variability and adaptive strategies utilised by both women and men among pastoral communities in Uganda. The results show that climate change and variability increases women's workload as men migrate in search of alternative income-generating activities. In addition, gender is shown to be a factor in the choice of a number of coping and adaptive strategies. Women, in particular, were found to have low coping and adaptive capacity due to the inequality in access to and control over resources.

Chapter Nine focuses on the impacts of climate change and variability and the different traditional coping mechanisms employed by pastoralist women in Somali Region of eastern Ethiopia. The author examines the impacts of climate change within the context of Somali culture before delving into the various indigenous adaptation mechanisms and complex adaptive strategies utilised by the women.

"Traditional Adaptation Mechanisms to Climate Change and Variability among Women Pastoralists in South-western Zimbabwe" in the last chapter presents the long-established adaptation mechanisms to climate change and variability used by women pastoralists in south-western Zimbabwe. Utilising the history of pastoralism in south-western Zimbabwe as the point of departure, the authors present the manifestation of climate change in the area and examine how the change affects women's roles in the pastoral economy, together with the coping and adaptation mechanisms the women employ in response to climate change.

Notes[1]

The Kyoto Protocol aims to attain quantified "limitation or reduction of emissions of greenhouse gases not controlled by the Montreal Protocol from aviation and marine bunker fuels, working through the International Civil Aviation Organisation and the International Maritime Organisation, respectively" (United Nations 1998, 2). It requires signatories to implement and/or further elaborate policies and measures in accordance with their national circumstances. Among others, it recommends policies and practices for the enhancement of energy efficiency in relevant sectors of the national economy; the promotion of sustainable forms of agriculture in light of climate change considerations; and research on promotion, development and increased use of new and renewable forms of energy, of carbon dioxide sequestration technologies, and of advanced and innovative environmentally sound technologies, among many other steps; it also advocates making reforms in relevant sectors aimed at promoting policies and measures which limit or reduce emissions of greenhouse gases (*Ibid.*).

References

Aklilu, Amsalu and Alebachew, Adem. 2009. Assessment of climate change-induced hazards, impacts and responses in the southern lowlands of Ethiopia. *FSS Research Report* No. 4. Addis Ababa: Forum for Social Studies.

African Ministerial Conference on the Environment (AMCEN). 2011. *Addressing climate change challenges in Africa: A practical guide towards sustainable development*. Nairobi: AMCEN Secretariat.

Anderson, S., Hesse, C. and Nassef, M. 2009. *Pastoralism and climate change: Enabling adaptive capacity*. London: ODI.

Cannon, T. 2002. Gender and climate change hazard. *Gender and Development*, 10 (2), 45-50.

Conroy, C. 2002. *Socio-economic methodologies for natural resources research best practice Guidelines*. London: Natural Resource Institute of Greenwich.

Dankelman, I. 2002. Climate change: Learning from gender analysis and women's experiences of organizing for sustainable development. *Gender and Development* 10 (2), 21-29.

Denton, F. 2002. Climate change vulnerability, impacts, and adaptation: Why does gender matter? *Gender and Development* 10: 10-20.

Djoudi, H. and Brockhaus, M. 2011. Is adaptation to climate change gender neutral? Lessons from communities dependent on livestock and forests in northern Mali. *International Forestry Review* Vol.13 (2),123-134.

Environmental Protection Agency (EPA). n.d. *Climate change basics*. Available online <http://www.epa.gov/climatechange/basics/>. Retrieved on 11th August 2012.

Hesse, C. and L. Cotula. 2006. *Climate change and pastoralists: Investing in people to respond to adversity*. Sustainable Development Opinion. London: IIED.

Inter-governmental Panel on Climate Change (IPCC). 2001. Summary for Policy Makers. In *Climate change 2001: Impacts, adaptation and vulnerability*. Intergovernmental Panel on Climate Change. Cambridge: Cambridge University Press.

——.2007. *Climate change 2007: The fourth assessment report of the Intergovernmental Panel on Climate Change*. Cambridge: Cambridge University Press.

IPCC TAR. 2001a. *Climate change 2001: The scientific basis*. IPCC Third Assessment Report. London: Cambridge University Press.

——.2001b. *Climate change 2001: Impacts, adaptation and vulnerability*. IPCC Third Assessment Report. London: Cambridge University Press.

Kraub, J. 2011. *Gender and climate change: Gender experiences from climate related GIZ projects*. Bonn and Eschborn: GIZ.

Levina, E. and Tirpak, D. 2006. *Adaptation to climate change: Key terms*. Paris: Organisation for Economic Co-operation and Development (OECD).

Mogotsi, K., Nyangito M.M. and Nyariki D.M. 2011. The perfect drought? Constraints limiting Kalahari agro-pastoral communities from coping and adapting. *African Journal of Environmental Science and Technology* 5(3), pp. 168-177.

Neumayer, E. and Plümper, T. 2007. The gendered nature of natural disasters: The impact of catastrophic events on the gender gap in life expectancy, 1981–2002. *Annals of the Association of American Geographers*, 97 (3), pp.551-566.

Oesterle, M. 2008.. From cattle to goats: The transformation of East Africa Pokot pastoralism in Kenya. *Nomadic Peoples* 12, 81-91.

Organisation for Economic Co-operation and Development (OECD). 2008. *OECD environmental outlook to 2030*. Paris, France: OECD.

Oxfam International. 2005. *The Asian Tsunami: The challenge after the Jakarta Summit.* London: Oxfam International.

----.2008. Survival of the fittest. Pastoralism and climate change in East Africa. *Oxfam Briefing Paper 116*. London: Oxfam International.

United Nations (UN). 1998. *Kyoto Protocol to the United Nations framework convention on climate change* New York: United Nations.

United Nations Framework Convention on Climate Change (UNFCCC). 2007. Climate change: Impacts, vulnerabilities and adaptation on developing countries. Bonn: United Nations Framework Convention on Climate Change.

----.2009. Copenhagen accord. Draft decisions of the fifteenth session of the Conference of the Parties held in Copenhagen, Denmark, December 7- 18, 2009. Bonn: United Nations Framework Convention on Climate Change.

----.2010. The Cancun Agreements. Bonn: United Nations Framework Convention on Climate Change.

----.2012. *The Durban platform for enhanced action.* Report of the Conference of the Parties on its seventeenth session, held in Durban, South Africa, 28 November -11 December 2011. Bonn: United Nations Framework Convention on Climate Change.

Warren, R. 2006. Impacts of global climate change at different annual mean global temperature increases. In Schellnhuber, H. J. *et al.* (eds.), *Avoiding dangerous climate change*. London: Cambridge University Press.

World Bank (The). 2008a. *Biodiversity, climate change and adaptation: Nature-based solutions from the World Bank portfolio.* New York: The World Bank.

----.2008b. *Climate change adaptation and mitigation in development programmes: A practical guide.* New York: The World Bank.

Ziervogel, G. and Ericksen, P. 2010. Adapting to climate change to sustain food security. *WIREs Climate Change* 1, 525-540.

Chapter 2

Climate Change Impacts and Local Coping Strategies among Pastoral Women: Cases from the Southern Lowlands of Ethiopia

Aklilu Amsalu and Desalegn Wana[*]

Abstract

The impacts of climate change are unevenly distributed among different regions and socio-economic groups of a given society. Particularly, women in pastoral communities are the most affected by adverse impacts of climate change; yet, the gender dimensions of climate change have barely received research and policy attention. The objective of this research is to investigate the impacts of climate change on pastoral women and existing coping strategies among pastoral communities of Borena and South Omo zones in the southern lowlands of Ethiopia.

Data were collected from the field by employing PRA tools, such as key informant interviews, focus group discussions, expert interviews, and field observations. In addition, sectoral policies and strategy papers have also been consulted to augment primary data. The results indicate that existing socio-cultural situations of the pastoral community have put women in a disadvantaged position primarily because of the asymmetrical power relationship between men and women over control of resources and decisions related to the utilisation of household resources. In addition, climate change has compounded an already existing work burden on pastoral

[*]amsalu.aklilu@gmail.com, deswana2002@yahoo.com

women through increased distance of fetching water, firewood collection, fodder for livestock, and other related productive and reproductive roles. Crop cultivation, which has not been common practice among the pastoral communities of Borena and South Omo, has increasingly become women's regular activity. Institutional responses are generally immediate fixes to the problems than dealing with long term underlying factors that aggravate the impacts on pastoral women. Future interventions need to focus more on women empowerment, family planning and reproductive health, girls' education, and the adverse impacts of harmful socio-cultural practices that complicate the impacts of climate change on pastoral women in the area.

Keywords: *Climate change, local adaptation, pastoral women, Ethiopia*

Introduction

There exists mounting evidence that human activities are forcing climate change processes (IPCC 2007). The increase in global population and economic growth, mainly increased production and consumption of fossil fuels, expansion of agricultural lands and deforestation have resulted in increased concentrations of greenhouse gases in the atmosphere (OECD 2008). The consequences of climate change are multi-faceted such as health-related risks, natural disasters, hunger and loss of livelihoods. Although climate change is generally considered as a global scale problem, the distribution of its consequences is uneven across different regions, ecosystems and various sections of society.

Climate change presents insurmountable challenges to human society in the 21st century. Africa is largely considered as one of the most vulnerable continents to the impacts of especially climate variability (Stige *et al.* 2006; IPCC 2007). Due to global warming, Africa is expected to face more variable climate and often frequent and severe weather events which will result in episodes of disease outbreaks, hunger and displacement (IPCC 2007). Similarly, the dryland ecosystems, where frequent drought and rainfall variability are defining characteristics, may severely be affected by climate change impacts. Lambrou and Piana (2006, 5, cited in Brody, Demetriades and Esplen 2008) succinctly put it as that "the vulnerability or susceptibility of a population group to the effects of climate change depends on the resilience of the surrounding natural landscape unit and society's capacity to adapt".

The pastoral livelihoods, which are mainly constituted of the delicate balance among pasture availability, livestock production and human demography, are believed to be highly resilient to periodic shocks of drought and rainfall variability (Hesse and Cotula 2006). However, the recent climate change via increased frequencies of extreme events such as drought and flooding undermines the pastoral livelihood systems by destabilising such a balance of pasture (for example bush encroachment and invasive species), livestock and people, and may consequently result in destabilisation of the overall production system and loss of livelihoods (Aklilu & Alebachew 2009). The spatial distribution and availability of pasture and water – the two most critical resources of the pastoral livelihoods – are highly dependent on the pattern and availability of rainfall. This rainfall unreliability is mainly caused by the climatic forcing factors such as *El Niño*, Southern Oscillation (ENSO) and North Atlantic Oscillation (NAO). Due to such forcing mechanisms, the amount of rainfall becomes lower and often the timing of rainfall turns out to be more erratic (Stige *et al.* 2006).

Even in the pastoral production systems, the impacts of climate change are not equally distributed among different segments of society. Some of the existing studies (Nelson *et al.* 2002; UNDP-GEF 2010) report that climate change has a gendered dimension and women tend to be highly vulnerable to climate change-related disasters and food insecurities. Women are invariably the most important contributors of livelihoods in pastoral livelihood activities, but have limited access to and control over vital resources such as land, livestock and household income. This socio-cultural 'discrimination' has put women in a disadvantaged position because of the asymmetrical power relationship between men and women over control of resources and decisions related to community and household resources. The recent climate change may have an additional burden over women in dryland ecosystems through increased distance of fetching water, fuel wood collection, fodder for livestock and other related productive and reproductive roles (Nelson *et al.* 2002). Despite such a pivotal role of women in the livelihood activities and their being the most likely to be affected by the adverse impacts of climate change, there has been very little detailed investigation on the multi-faceted impacts of climate change on women and how such gendered dimension of climate change in dryland ecosystems affect the adaptation to climate change impacts. More explicitly, the gender dimension of climate change impacts has largely been neglected in the design and implementation of national and regional adaptation policies and strategies.

Ethiopia is one of the poorest nations of the world; vulnerable to the impacts of climate variability and change. High dependence on natural resources and climate sensitive livelihoods, coupled with the existence of rampant poverty and variable weather events put the country in the most vulnerable position (Conway and Schipper 2011; Oates, Conway and Calow 2011). The country's environmental resources, including water, forests and range resources, and major socio-economic sectors (agriculture and health included), are vulnerable to climate change and variability, with huge impacts on the local ecosystem and the socio-economic well-being of the population (Arndt, Robinson and Willenbockel 2011). The limited economic, institutional and logistical capacity to mitigate and adapt to climate change exacerbates the vulnerability of many people and communities to climate hazards that are occurring through global climatic changes and land use/cover transformations.

Pastoral communities in Ethiopia constitute about 12 per cent of the total population of 77 million (CSA 2008). Over the past several decades, the country has been repeatedly hit by droughts, famine and epidemics that relate to changing climatic conditions. Pastoral communities have been particularly affected by these problems that have been occurring recurrently. These people are among the most resource-deprived and the most geographically and politically marginalised. There have been studies conducted on climate change-related problems, impacts and responses among the pastoral communities of Ethiopia (Aklilu and Alebachew 2009; HoA-REC and GTZ 2010). However, the gender aspects of climate change and the differential impacts on pastoral women in the country have not been adequately addressed. This research is aimed at investigating the impacts of climate change on pastoral women and existing coping strategies among pastoral communities of Borena and

South Omo zones in the southern lowlands of Ethiopia. Its specific objectives are:

- to examine the differential impacts of climate change on pastoral women;
- to assess the extent of vulnerability of pastoral women to future impacts of climate change;
- to analyse community responses focusing on the specific roles of pastoral women in coping with the impacts of climate change; and
- toappraise local and institutional responses to the impacts of climate change.

The significance of this study will be two-fold:

Firstly, it helps to fill in the knowledge gap on the gendered dimensions of climate change impacts on pastoral communities in general and women in particular in Ethiopia. In a way this research documents the pivotal role pastoral women play in the livelihoods of the pastoral production systems and how they are being affected by the impacts of climate change-related disasters and risks. Moreover, it highlights on the gender differences not only on the impacts of climate change but also different responses and coping strategies to climate change impacts and related disaster risks in the pastoral communities.

Secondly, the findings of this study will inform local and regional adaptation policy makers of the differential impacts of climate change on pastoral women and varied responses to climate change-related disaster risks in the pastoral regions of Ethiopia.

The Gender Dimensions of Climate Change Impacts: An Overview

Climate change is now posing tremendous challenges for the global community. In particular, global warming has resulted in the increase of extreme weather events such as tropical cyclones, floods, droughts and heavy precipitations in different parts of the globe (UNFCCC 2007). Melting of snow packs in the high latitudes, sea level rise along the coast, contraction of glaciers at high mountains in the tropics and increased severity and frequencies of extreme weather events are peculiar characteristics of climate change in the tropics and sub-tropics (OECD 2008). The manifestations of climate change are found across every corner of the globe. No part of the globe or society can, therefore, be immune from the pervasive impacts of climate change. However, the type and extent of the impacts could vary for different socio-economic and ecological contexts. For instance, flooding and sea level rise, both of which are extreme events, destroy property and human lives indiscriminately while the effects of droughts and diseases related to climate change might have differential impacts, hence, the concern to address climate change impacts by considering differential impacts is appealing.

Many recent scientific works reveal that climate change impacts are disproportionally distributed among different regions, ecosystems, gender, age and the level of economic development (IPCC 2007; UNFCCC 2007). Developing countries contribute very little in terms of emitting climate change-driving greenhouse gases but suffer enormously from climate change related risks and hazards. This is so mainly because these countries are highly dependent on climate-sensitive economic activities, mainly agriculture, for their livelihoods and incomes, and thus the projected impacts would be more severe. The impacts of climate change will manifest in the loss of productive land and failure of crop production, and decline in yields which would eventually result in increased food insecurity and vulnerability of the rural communities in developing countries of the tropics and sub-tropics (Ziervogel and Ericksen 2010). Africa, due to its location in the tropics and high level of endemic poverty, is under great stress due to climate change impacts (UNFCCC 2007). African countries usually suffer from endemic poverty, burden of tropical vector-borne diseases, intermittent conflicts and HIV/AIDS. Climate change impacts in various forms are thus exacerbating the existing problems, such as by spreading vector-borne diseases and crop pests to previously unexposed areas, which in turn overburdens the already over-stretched health systems, weak economy, and endangers fragile livelihood structures.

Within Africa itself, climate change impacts also vary. For instance, dryland ecosystems are one of the most vulnerable victims of climate change. Drylands are usually characterised by moisture stress, unreliable rainfall and fragile landscapes. The livelihood systems of drylands are mainly dependent on cattle-rearing and nomadic pastoral way of life, which is generally believed to be resilient to periodic shocks of drought and shortage of pasture (Brody Demetriades & Esplen 2008). The widely held view is that pastoral production system as a way of life and culture as such, thought to be "long-lasting cultural adaptation to their environment were now considered as vulnerable, ecologically risky and non-sustainable" (Oesterle 2008, 81). Obviously, the increasing frequency of drought and heavy precipitation have affected availability of grass and pasture and put the whole livelihood system at risk. The presumed 'equilibrium' between cattle production, human population and pasture availability is already disrupted by climate change impacts (Hesse and Cotula 2006).

Recent studies in East Africa (such as Oesterle 2008) have reported that, due to human population growth, the grazing ecosystem of the Pokot savannah in north-western Kenya cannot support large numbers of cattle, which has forced them to look for other sources of supporting livelihoods. It is generally the surge in human population which leads to the increase of cattle population and fosters overgrazing and eventually results in the degradation of the grazing land. Over-grazing due to over-stocking of the range ecosystems in the drylands, particularly in the Sahel and East Africa, causes trampling and desiccation of the soil during drought periods and this eventually leads to flash flooding when the rain falls. Drought and flooding, which are extreme events of climate change in drylands, reinforce each other and probably become one of the most difficult challenges for pastoral communities and governments to deal with. The question whether drylands are more vulnerable to climate change impacts as compared to other ecosystems is not disputable. However, whether climate change exclusively explains the adverse impacts on pastoral livelihoods and resources or it plays a compounding role to the already prevailing demographic and livestock pressure (Malthusian perspective of people-environment relationships in such environments) remains to be explored. Huq, Reid and Murray (2006) argue that unsustainable development is the underlying cause of climate change, and development pathways will determine the degree to which social systems are vulnerable to climate change.

Turning to the subject of whether climate change affects women differently from men, suffice to say that there exists mounting evidence (Denton 2002; IPCC 2007; UNFCCC 2007) that women suffer more from climate change-related problems than men do. A number of issues underscore the differential impacts of climate change on women, particularly in pastoral communities.

Firstly, women and men play different social roles and climate change is expected to have differential impacts. Women usually take the role of caring for children and weak livestock. When severe drought occurs, men migrate in search of water and pasture for the cattle to distant places, leaving children and women behind. This leaves the women with the burden of taking care of children, feeding cattle and the family single-handedly.

Secondly, women are highly dependent on the natural environment because they are involved in the procurement of food for the family. As such, any degradation in the environment would adversely affect women more than men. In this regard, some feminist scholars such as Esther Boserup (1989, cited in Dankelman 2002, 23) argue that, "it is women-the-gatherer not men-the-hunter who were traditionally a source of sustainable food supply." As well, the United Nations Framework Convention on Climate Change (UNFCCC) (2007) revealed that increased frequencies of drought and associated heavy precipitation may undermine the availability of food for the family and consequently increase the burdens on women. As Dankelman (2002, 23) aptly puts it, "every down brings with it a long march in search of fuel, fodder and water". These critical resources are becoming scarce and the long march is becoming longer and even may reach to the limit beyond women's physical endurance (Dankelman 2002).

Finally, climate change has adverse impacts on women because of the cultural burdens in many rural societies in which women have no access to and control over productive assets such as livestock and land (Röhr 2007). Such situations make pastoral women more vulnerable to the adverse impacts of climate change. It is, therefore, not surprising that gender also plays a role in relation to climate change. Yet, the topic has not been explored sufficiently, and many people still find it difficult to comprehend the ways in which gender might be a factor in climate change, or how it should be addressed (Röhr 2007; Aklilu and Alebachew 2009).

Methodology

The Study Area

The study was carried out in two zones of the southern Ethiopian lowlands, namely Borena and South Omo, where the impacts of climate change is pronounced. Borena zone is located within the Oromia Regional State while South Omo is located in the Southern Nations, Nationalities and People's Region – SNNPR (see Figure 1). Borena is one of the 17 administrative zones in the Oromia Region and is divided into 10 *woredas* (districts). It covers a total area of 48,743 km^2 and has a total population of 966,467 (CSA 2008). Yabello is the capital of Borena Zone.

South Omo is one of the 9 administrative zones in the SNNPR. The zone is named after Omo River which descends from the central highlands of the country. South Omo covers an area of 24,378 km^2 and is divided into 8 *woredas*. Jinka town is the capital of South Omo Zone. The zone has a total population of 577,000 (CSA 2008). It should be noted that females account for about half of the total population in the two zones. Table 1 presents some demographic data of the two zones.

Table 1: **Population and area size of Borena and South Omo zones**

Zone	**Population ('000)**			**Total area ('000km²)**	**Density (persons/ km²)**
	Male	Female	Total		
Borena	489	477	966	48.7	20
South Omo	288	289	577	24.3	24

Source: CSA (2008)

The average annual rainfall in the area ranges from 400 to 700mm, with wide temporal and spatial variations (Aklilu and Alebachew 2009). The mean annual temperature is 25°C. The rainfall pattern is erratic, characterised by late onset, early cessation, and sometimes complete failure. Analysis of the annual rainfall for all the three stations varies widely (*Ibid.*). Similarly, the long term pattern of the mean annual temperature in the area clearly shows an increasing trend. It is well recognised that small increases in temperature can result in measurable impacts on the health of human beings and livestock as well as the availability of water, food and feed resources. The area experiences a semi-arid climate with double wet and dry seasons. Hence, rainfall distribution is bimodal. The main rainy season, which is locally called *ganna,* lasts from March to May and accounts for about 60 per cent of the total rainfall in the area. The short rainy season *(hagayya)*, which is from September to October, is a very important rainy season in the area. Similarly, there are two dry seasons: the long dry season called *bonna,* lasting from November to February, and the short dry season from June to August.

Pastoralism is a way of life and a predominant source of livelihood in the study areas. Cattle and goats constitute the major livestock types kept by local communities. Basic physical and socio-economic infrastructures in the two zones are very weak and, for the most part, lacking.

Figure 1: **Location of the Study Areas**

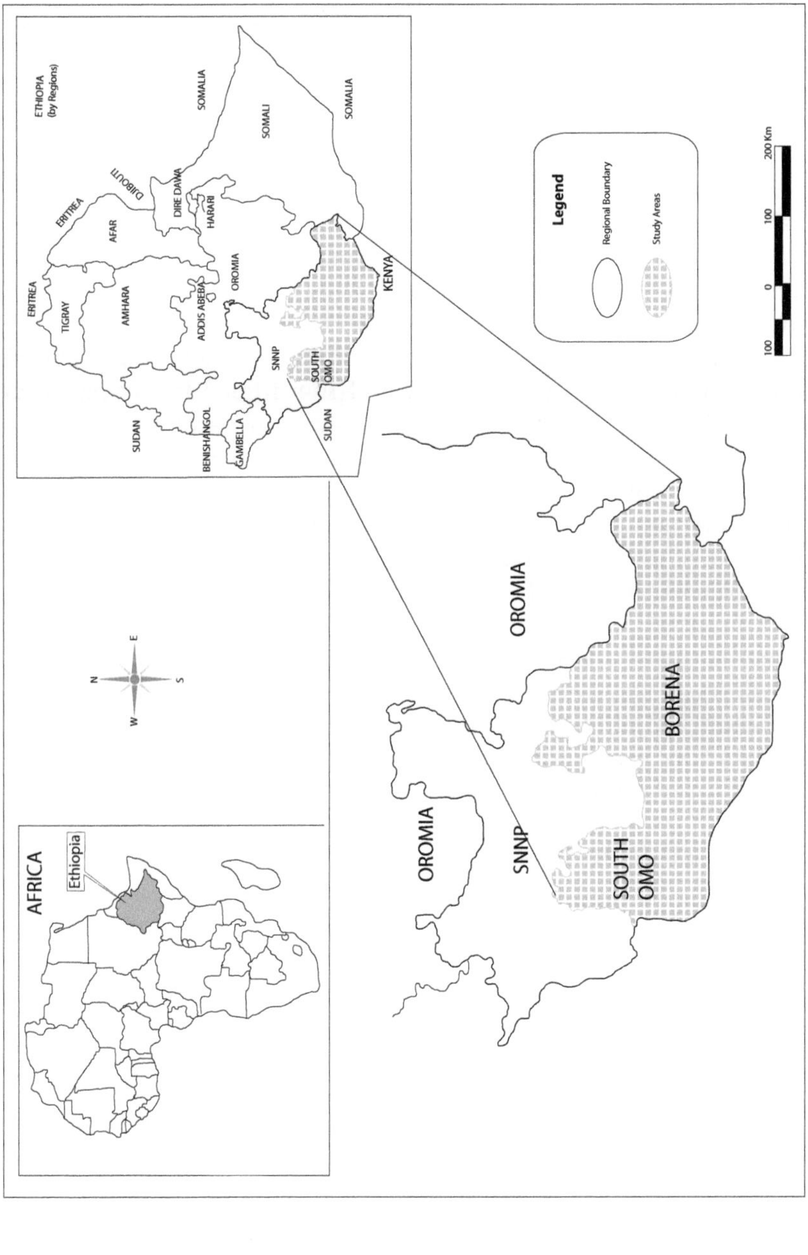

Data Sources and Methods of Analysis

This study used both primary and secondary sources of data. Primary data were collected from the field through Participatory Rural Appraisal (PRA) tools, such as Key Informant Interviews (KIIs), Focus Group Discussions (FGDs), Expert Interviews (EIs), and Field Observation (FO). The fieldwork was conducted in October 2011. Field notes of previous study (2008) in the area by the first author of this manuscript were also consulted. Key Informant Interviews were held with knowledgeable individuals selected from the community. Focus Group Discussions were held with separate groups of men and women in each *woreda* of both study areas to elicit information on climate change-related problems, the impacts on pastoral livelihoods and coping strategies, and the differential impacts on women. Three FGDs from each study areas were conducted. The composition of the FGD participants took into account the inclusion of pastoralists of different age groups and experience. Four of the six FGDs were gender-separate while the remaining two were mixed groups. In addition, EIs were carried out with experts from relevant sector offices, such as the zonal and *woreda* offices of agriculture, women's affairs offices, and non-governmental organisations of the respective study areas.

Secondary data were obtained from review of related literature and analyses of relevant policy documents. The review was facilitated by a checklist of major issues surrounding the gender dimensions of climate change impacts and coping strategies.

Results and Discussion

Climate change is affecting pastoral livelihoods in Africa, in general, and Ethiopia, in particular. The southern lowlands are among the most vulnerable regions to the effects of climate change in the country. Climate data for the southern lowlands indicate rising temperatures and erratic and unpredictable rainfall (Aklilu and Alebachew 2009). Extreme weather conditions are likely to worsen existing water shortages as these areas will tend to become drier. This section discusses climate change-related problems and their impacts on pastoral communities, their distinct impacts on pastoral women, local coping strategies, and institutional and policy responses.

Climate Change-Related Problems and Impacts on Pastoral Communities

Climate change is one of the major constraints affecting the livelihoods of pastoral communities in the southern lowlands of Ethiopia. The area has increasingly been exposed to several climate change-related problems, the most prominent ones of which include droughts, flooding, diseases and pests.

Drought is not a new phenomenon to pastoral communities of the southern lowlands of Ethiopia. The study area has repeatedly been affected by severe droughts over the past years. Previously, droughts used to occur every 5 to 8 years in the area. However, the drought cycle has changed in recent times and become more frequent, occurring every 2 to 3 years (Aklilu and Alebachew 2009). This has also been noted during focus group discussions and expert interviews. Particularly, the last ten years have seen more frequent and prolonged droughts. Repeated prevalence gives limited or no time to recover from the impacts of the preceding drought. For instance, droughts prevailed in the years 1998/99, 2000/01, 2003/04, 2006/7, 2007/8, and 2010/11 in the study area and caused widespread malnutrition, excessive loss of livestock, and massive migration. Moreover, the rains were not only insufficient in amount but also extremely unpredictable to plan and properly carry out livelihood activities. In 2010/11, there was no rainfall at all for over 15 months in the area. Longer and more frequent droughts are likely to result in a significant rise in destitution among pastoral groups since such extreme events decimate herds and prevent their reconstitution (Hesse and Cotula 2006).

Local communities have been affected by recurrent droughts that hit the area every 2 - 3 years (Tiki, Oba and Tvedt 2011). The droughts have increased aridity and the disappearance of some important indigenous tree species with medicinal and nutritional value for humans and livestock. In Borena, traditional deep-water wells (*ellas*) are normally reserved for dry season and used from December onwards. But as of recent, this is changing, and the traditional wells are being used earlier on starting from July due to severe droughts in the area. This is gradually leading to the degradation of the traditional wells.

Participants in FGDs and *woreda* experts have indicated that the recurrent and extended droughts, coupled with existing and newly emerged livestock diseases, are causing more illness and livestock deaths. For example, camels and goats that are considered most resistant to drought are severely affected by newly emerged diseases. On the other hand, households are forced to move their livestock to distant places to which they have never been exposed before during intense droughts, potentially exposing their herds to different environments with health risks. Tick and skin diseases on camels, cattle, goats and sheep are pervasive problems during intense droughts.

Loss of livestock is a serious challenge threatening livelihood of the people in the area. Over the past two decades, the number of livestock per household has fallen drastically because of deaths caused mainly by diseases and under-feeding. For instance, in Borena Zone, livestock holding per household has declined from 10 to 3 oxen, 35 to 7 cows, and 33 to 6 goats; and in South Omo, from 30 to 21 cows, 38 to 23 goats and 36 to 21 sheep (Aklilu and Alebachew 2009). According to informants, declining livestock productivity and reproductive capacity is another serious threat they are facing. The loss of important household resources such as livestock and cropland has led to increased dependence on external assistance, for both food and non-food items. Poorly developed rural markets that have weak links to facilitate the exchange of livestock with food crops have been mentioned by the community as main problems restraining their capacity to cope with the disasters.

The southern lowlands have previously been rich with rangeland resources. However, the rangelands have gradually been degraded to sufficiently support livestock population in the area. Valuable range resources are being replaced by unpalatable bushes and low quality grass species. In particular, bush encroachment is rampant in the area. Declining rangelands and crop productivity is reducing the amount and quality of already scarce forages and crop by-products that are used to feed livestock.

In addition to droughts, the risk of flooding is also rampant in the study area. In particular, South Omo Zone is one of the places most affected by flooding hazards in the southern Ethiopian lowlands. Flash floods and floods from Omo and Woito rivers have caused widespread destruction of life and basic household assets. Although flooding is common to the area and the people benefit from it, the pattern in recent times has changed and become detrimental to the people and their resources. The floods have become frequent, out of season or erratic, and sudden. For instance, in August 2006, flooding caused the death of 364 people and about 2,700 livestock, and the displacement of over 15,000 people (*Ibid.*). Sudden and very intense flash and mud floods of wider spatial coverage were observed all over the study area during the fieldwork in October 2011. Describing the flooding incidence, a key informant in South Omo aptly put it: "In previous times heavy rains did not cause much flooding but now a day's even little rain yields very intense flash floods". This clearly indicates the disruption of the delicate balance of human demography, pasture availability and livestock density which generally characterise the pastoral production system (Hesse and Cotula 2006; Brody, Demetriades & Esplen 2008; Oesterle 2008). According to informants, the flash floods have also a wide range of impacts such as destruction of crop farms, livestock, infrastructure, and often loss of human lives. Apart from the direct loss of life and property, the floods have led to the outbreak of fatal diseases such as acute watery diarrhoea (AWA), cholera, fever, and malaria. Similar cases have also been reported in Bangladesh and the Caribbean (Dankelman 2002). Furthermore, the floods have hindered land preparation activities for the next cropping season in South Omo where riverside crop cultivation is common. In the summer of 2008, major zonal towns, like Yabello and Jinka were hit by unusual flash floods that caused considerable loss of property and basic infrastructure (Aklilu and Alebachew 2009).

Climate change has direct and indirect impacts on the prevalence and spread of diseases and pests (UNFCCC 2007). The incidences and distribution of diseases and pests in the study areas have changed over the past few years. Locally prevalent diseases are expanding and new types, some of which have not yet been identified, are emerging. Occasional outbreaks of epidemic diseases, such as malaria, AWD, measles, Rift Valley Fever, and cholera are becoming increasingly common in the area. Even in places where malaria was rare, the disease has become severe and deadly. Furthermore, as the respondents indicated, new diseases, which include bouts of cold, and respiratory and intestinal infections, are becoming more frequent. Moreover, water-borne diseases that cause livestock and human deaths are also prevalent due to occasional floods.

Impacts on Pastoral Women

Although climate change generally affects pastoral communities, the impact on women has become more pronounced. The fact that pastoral women shoulder multiple household responsibilities makes them more susceptible to and affected by the impacts of climate change-related problems. Box 1 presents the responsibilities of pastoral women in the study areas as outlined during FGD, both by women and men groups.

Box 1: **Household responsibilities of pastoral women**

The main household responsibilities of pastoral women include:

- Fetching water, from wherever available, for household use;
- Food preparation, which includes grinding grain and collecting wild food sources;
- Taking care of children, including feeding and looking after them;
- Firewood collection;
- Milking livestock (only in Borena);
- Gain procurement from the market for household consumption;
- Fetching water and collecting grass for weak livestock, calves and small ruminants;
- Crop cultivation and cropland management such as sowing, weeding and hoeing (mainly in South Omo); and
- Construction of huts (only in Borena).

As outlined above, pastoral women in the study areas have several household responsibilities: taking care of children, managing young calves, looking after sick and weak livestock, milking livestock, fetching water and collecting firewood, and household food preparation. In addition to such household chores, construction of huts is the sole responsibility of women in Borena. Added to this, climate change-induced hazards create additional burdens on women in many ways and make them more vulnerable to its impacts (Denton 2002; Dankelman 2002). According to Aklilu and Alebachew (2009), a significant proportion (about 80 per cent of the respondents) of the pastoral households in the southern lowlands of Ethiopia reported that women are more affected than men by the impacts of climate change-related problems. Particularly, during drought years, the workload on women increases to meet household responsibilities. For instance, in Borena Zone, women have to spend on average 4 to 6 hours a day fetching water and fodder for small ruminants and calves. In South Omo, the burden on women is even greater since the responsibility of feeding family members largely rests on them, which they carry out by collecting wild fruits, plants, berries and edible roots.

As such, increased workload and resource shortages have put pastoral women in the study area under pressure to meet household food requirements. Cooking, including procurement of grain from the market, collecting wild fruits and leaves for household consumption, is the sole responsibility of women. In times of food shortage in the households, women have to first feed all members of the household and are the last to get food – if there is any leftover. This exposes women to malnutrition and related health problems, with subsequent impacts on children and the entire household. On the other hand, when women fail to meet household responsibilities, they get into disputes with husbands that strain intra-household relations. In extreme situations, such disputes often lead to divorce and eventually to destitution of pastoral women (Interview with the Head of Women's Affairs Office in Borena, October 2011).

In order to overcome the problem of drought-induced feed and water scarcity, pastoral households move their livestock to places where they find these resources. Fetching water for young livestock and the family is a daunting task of women especially during drought seasons.

At times of conflict and violence, pastoral women also suffer from other negative consequences. For instance, in the absence of men, they shoulder the sole responsibility of sustaining the household, which increases their workload. Many women and young girls also fear going out of their villages, as they are often the targets of attacks, and this makes it difficult for them to fetch water, collect firewood or go to the market. Besides, when part of the family is moving, usually women and children remain behind. Although women shoulder the family responsibility they cannot take critical decisions on the remaining household resources, mainly livestock. To illustrate, pastoral women in the study area are not allowed to sell livestock even in times of extreme adversities. In line with this, Röhr (2007) points out that lack of access to and control over household resources makes women more vulnerable to climate change impacts.

In relation to climate variability and change, competition over scarce pasture and water resources among pastoralists increases the risk of conflicts (Bouch and Mammo 2008). The conflicts usually occur during severe droughts, following pastoral mobility in search of water and pasture for livestock. For instance, the violent conflict between Borena and Konso in March 2008 led to the displacement of 27,000 people and raiding and looting of 1,500 heads of livestock (Aklilu and Alebachew 2009). The conflicts often cause social and economic instability in which children are forced to drop out of school; farms and settlements are left unattended; and crops may remain unharvested. Although the conflicts are usually sparked by competition over scarce resources, they often comprise ethnic dimensions which complicate the impacts of climate change on pastoral communities. All these conflict-related problems and disruptions bring about increased work burden on pastoral women in terms of their household responsibilities.

Local Coping Mechanisms

Local communities use a range of measures to cope with the impacts of climate change-related problems (Girum, Degefa and Getachew 2008; Smucker and Wisner 2008). Herd diversification, crop cultivation, social support mechanisms, seasonal mobility and resource management strategies are some of the most commonly used strategies in the study areas. Cutting back on the number of meals per day, use of wild fruits and cattle blood are also additional strategies used to deal with intermittent food shortages. However, their capacity to cope with the impacts of the disasters is limited. On the other hand, some of the other strategies, such as cultivation of unsuitable areas, overgrazing, charcoal burning and fuel-wood selling are not only unsustainable but would also cause further degradation of resources and desertification.

However, herd diversification has become an important coping mechanism in the study areas. Diversification offers sustained supply of various livestock products allows the manipulation of different ecological potentials of livestock and helps to spread risks associated with changing climatic conditions. In general, there appears to be a shift from the production of grazers (cattle and sheep) to browsers (camels and goats). Camels and goats are relatively more adaptive to changing climatic conditions and vegetation cover. Hence, the degradation of pasture and bush encroachment has forced pastoralists to diversify their livestock types. In the study area, camels are a recent introduction especially promoted by non-governmental organisations (FARM-Africa in South Omo, and SOS-Sahel and Action for Development – AFD in Borena). Camels and goats give milk even during dry periods, when lactating cows can hardly be milked. Furthermore, the timing of milk availability is vital for households as goats can ensure continuous supply. Herd and household mobility is a flexible strategy to optimise the utilisation of natural resources and a response to disaster risks, such as disease outbreaks, flooding risks, droughts and conflict. Although long distance mobility is a long-standing tradition of the pastoral communities, its duration and frequency has changed in response to increased climatic stresses.

Due to increased frequency and prolonged droughts, local communities reconstitute their livestock types and compositions. For instance, camels are introduced in the production system as a response to the changing ecology and feed availability.

Although keeping camels is an old tradition in Borena, pastoral households tend to gradually focus more on the production of camels. In South Omo, however, camel production is a new activity introduced by non-governmental organisations operating in the area. Pastoral communities in South Omo also focus on goats in response to recurrent droughts and expansion of bushes and shrubs. In general, there appears to be more reliance on browsing livestock than cattle as the rangeland degrades and is gradually replaced with bushes and trees. Such changes have also been observed among the pastoralists of East Pokot in Kenya (Oesterle 2008).

Pastoral households in the study area undertake crop farming as a complementary strategy to avoid potential risks of sole dependence on livestock production. Crop cultivation is spreading to rangelands and to areas that had never been put under cultivation, particularly in Borena. The expansion of crop cultivation is mainly linked to the decline in range resources as well as a decrease in both per household livestock size and productivity. The responsibility of farming and managing cropland largely rests on women. However, given the unreliability of rainfall, increased frequency of droughts and associated crop diseases and pests, the sustainability of rain-fed crop cultivation as a viable coping strategy is questionable. Although crop farming is being widely used as an *ad hoc* coping strategy by pastoral communities, increased dependency on it might increase vulnerability of the local community to future climate change-related risks. Moreover, crop cultivation on fragile rangelands would accelerate soil erosion by wind and water.

For cultural reasons, pastoral households prefer to keep as many livestock as possible since livestock number is socially valued. However, with severe droughts and other climate-related risks, households in the study area have increasingly come under pressure to sell or exchange their livestock. Although there are recent developments in improving market infrastructure, livestock marketing is generally weak in the area. According to survey results, about 78 per cent of the households in Borena and 33 per cent in South Omo have reported an increasing trend of livestock selling among pastoral communities in recent times (Aklilu and Alebachew 2009). Normally, households start responding to the problem by selling small ruminants, mainly goats and sheep. With increased intensity of the problems, however, households are forced to sell cattle that are the main sources of livelihoods and social prestige. In other words, pastoral communities sell their camels and cows when impacts of the problem get worse and other options have been exhausted.

Resource-sharing is a traditional social support mechanism practised by pastoral communities to support vulnerable members of the society (Smucker and Wisner 2008). The support mechanism ranges from simple sharing and provision of milk and other food items to contributing heads of livestock in order to re-stock the herds of affected families. Households who have received the support in turn contribute to the community by helping others through labour work and livestock herding. Local communities support relatives and other clan members by temporarily providing milking livestock (a goat or a cow) and other productive assets to overcome immediate household food shortages. However, the capacity of these social support systems has been weakened due to increased intensity of the problems.

Hence, pastoral communities are desperately looking for additional sources of income which include engagement in non-pastoral activities (*Ibid.*). Such activities include engaging in labour work in urban areas, charcoal-making, and firewood-selling. Women conduct petty trade and related income-generating activities.

During severe disasters, households are forced to change their normal food intake and adjust their consumption to the available household resources. Consumption smoothing usually involves adjusting diets to available food items, supplementing with edible wild plants and fruits, and reducing the amount of meals and the frequency of serving. Such adjustments involve reducing food intake from three times to once a day – giving priority to children and the elderly, and shifting from milk and milk products to more cereals and tea. Households in South Omo mix cattle blood with milk and water as main source of food. In Borena too, households mix milk with water and maize flour, and there is unprecedented consumption of tea, replacing milk during severe droughts.

Various types of resource management strategies are used by pastoral communities to deal with changing climatic conditions. Dynamic management of rangelands, water and livestock resources is widely practised in order to overcome over-grazing, allow fast regeneration of pasture, and in response to climatic variations within and between months in a year. Rangelands are traditionally managed by dividing into dry and wet season grazing considering water availability and herd size and structure. Particular areas, which are close to settlements and homesteads and considered rich in both pasture and water resources, are reserved for calves, milking cows, and weak livestock. Livestock are watered every second or third day, as opposed to daily watering, during the dry season in order to ensure proper forage-to-water intake. On the other hand, during the rainy season, ponds and deep wells are closed (mainly in Borena) and both livestock and humans use surface water instead (Aklilu and Alebachew 2009; Tiki, Oba and Tvedt 2011). Proper management and fair utilisation of rangeland and water resources is facilitated by traditional institutions (the *Gada* system in Borena and the guardians for grazing/cropland in South Omo). Pastoral communities practise controlled burning to allow regeneration of new pasture by eliminating unwanted weeds and control ticks, pests, rats and other disease causing organisms. Bush-burning is usually practised in September or October to take advantage of the short rains, which normally come in October. Furthermore, burning at this time assures effective control of ticks since grasses are dry enough to burn and effectively kill ticks. Although bush burning has been a long-standing resource management strategy, it was banned by the government since the 1970s as it was considered a harmful de-vegetation process.

Institutional and Policy Responses

The government of Ethiopia has adopted policies, strategies and action programmes aimed at poverty reduction, environmental protection and sustainable development. Although climate change is being mainstreamed in several sectoral policies and strategy documents, the country has no explicit policy on climate change yet (Aklilu and Alebachew 2009; Oates, Conway and Calow 2011). For instance, the Poverty Reduction Strategy Papers (PRSPs) offer little consideration to climate change. Rather, many of the proposed activities are directly aimed at reducing the impacts of climate change, mainly droughts (Conway and Schipper 2011). However, there have been institutional responses in various ways by state and non-state actors against the adverse impacts of climate change over the past few decades in the country. In the study area, government responses mainly focus on emergency assistance, asset protection and livelihood enhancement, disaster risk reduction, and conflict management and resolution.

The government of Ethiopia has long been engaged in supplying emergency aid to victims of various types of disasters. In the study area, the government has been supplying food and non-food items since the 1980s and particularly during recent drought and flooding. For example, during the 1984/85 and 1998/89 drought in Borena and the 2006 flood hazards in South Omo, the Federal and Regional Governments in collaboration with other partners distributed food, medicine and relief supplies (discussion with experts of Borena Zone Office of Agriculture and Rural Development). Rescue operations were carried out and tents, blankets, foodstuffs, mosquito nets and non-food items were distributed, and some health, sanitation and security services were provided. Rescue operations are followed by continued relief assistance and rehabilitation activities for some time in order to restore life and livelihood of the affected people. The Disaster Prevention and Preparedness Office at the district and sub-district levels work together with other sector offices and local NGOs at the initial stage of a hazard to study and report the situation to relevant authorities for immediate attention and support. During the FGDs, the participants reported that the government supports those critically food insecure and destitute pastoral households through its productive safety net programme. For instance, households in South Omo receive support through food-for-work and productive safety net arrangements. Livestock feed and water rationing has also been carried out in Borena Zone during severe droughts. However, emergency responses in most cases do not match the amount required to address emergency needs of the local community, nor delivered on time to local communities.

Based on the belief that protection of household and community assets reduces vulnerability to the impacts of climate variability and change, the Federal government introduced a Productive Safety Net Programme in 2004. The programme is aimed at supporting extremely vulnerable and destitute households through direct transfer of cash and food. It also mobilises the community through food-for-work arrangements to engage in soil and water conservation, bush clearing, and maintenance of rural roads. On the other hand, the government has been supporting households by providing different packages that facilitate the transition from an exclusively pastoral way of life to agro-pastoralism through the provision of seeds, fertilisers, tools and training. Nevertheless, such efforts should consider appropriateness of the local agro-ecology, socio-cultural fabric and other sustainability issues to practise crop cultivation and a sedentary way of life (Aklilu and Alebachew 2009).

In the study areas, there are Disaster Prevention and Preparedness units and committees in each *woreda* that are commissioned to provide early warning information, assess needs and monitor risks. However, most of these units suffer from poor organisation, limited capacity and resources (material and human), and a lack of networking with relevant sector offices. FGD participants, mainly in Borena, indicated that they are not satisfied with the government's responses particularly against droughts, diseases and pests, and conflicts that have been inflicting heavy losses. In fact, other sector offices undertake some development activities that enhance access to water (through wells and boreholes) and carry out environmental protection and rehabilitation programmes to improve range resources. Nevertheless, most of these activities are handicapped by material and capacity limitations and often a lack of continuity in order to enable households to cope with disasters.

Considering the complex nature and seriousness of conflicts, the government has set up various committees at different administrative levels. In this regard, there are encouraging efforts, particularly along with community policing, to identify and detect symptoms of conflict and tension, control crime and peacefully resolve disputes. This is indeed a hopeful direction. However, it should proactively engage and empower community institutions, and promote peaceful co-existence through public education, capacity-building and social development.

Over the past years, the NGOs operating in the study areas have been implementing several projects aimed at disaster impact mitigation and poverty reduction in pastoral communities. The projects have had considerable impacts, but also numerous problems in bringing about improved adaptive capabilities and securing livelihoods. The responses focus on disaster risk reduction, provision of social services, awareness creation, and conflict management and resolution. Disaster risk reduction through asset protection and livelihood diversification is basic intervention aimed at reducing vulnerability and enhance adaptive capacity of pastoral communities. Some NGOs in South Omo have projects and programmes aimed at managing droughts and other risks through the protection of household assets and community-owned natural resources and by enhancing livelihood sources. Specific interventions include water harvesting, identifying and fencing dry season grazing areas, de-stocking, use of early warning information, micro-credit services (for acquiring goats and camels), and emergency aid. The Women's Goat Project in South Omo is an exemplary intervention initiative designed to enhance the capacity of poor pastoral women by providing female goats to the women so that they are able to recover their stock. The impact of this project in reducing vulnerability of women to climate change impacts has been considerable, although its continuity to widen the coverage of beneficiaries appears to be limited.

Basic social services such as education, health, and infrastructure in the study areas are for the most part lacking and/ or poorly developed. Some NGOs are involved in the provision of education and health services to the community. However, the provision of these services is largely fragmented, less coordinated and short-lived. Social services that particularly focus on women empowerment and reducing their work burdens are quite limited. There are also NGOs which are actively engaged in peace-building activities. They operate by mobilising traditional leaders and elders, providing training, organising peace conferences, and delivering information, finance and materials for conflict resolution efforts in conflict-prone areas (Aklilu and Alebachew 2009). However, despite the wider scale and frequency of potential conflicts in the area, available interventions are quite limited and less effective to build sustainable peace.

Summary and Conclusion

Climate change has become one of the pressing challenges of human society. However, its impacts are disproportionally distributed among different regions and socio-economic groups of society. The consequences of climate change are multi-faceted, including natural disasters, hunger and loss of livelihoods and health-related risks. Pastoral communities in Ethiopia are among the most vulnerable groups to climate change impacts. However, the impacts are not equally distributed among different segments of the pastoral community. Women are particularly affected by climate change-related problems in various ways.

The findings of this study indicate that women have limited access to and control over productive resources such as land, livestock and household income. The existing socio-cultural situations of the pastoral community has put women in a disadvantaged position mainly because of the asymmetrical power relationship between men and women over control of resources and decisions related to the utilisation of household resources. In addition, climate change has compounded an already existing burden on pastoral women through increased distance they travel to fetch water, collect firewood, fodder for livestock, and their engagement in other related productive and reproductive roles. Crop cultivation, which had not been a common practice among the pastoral communities of Borena and South Omo, has now increasingly become a women's regular activity. Nevertheless, due to unreliable rains and flash floods there has been frequent crop failure and farm destruction that has caused labour wastage due to sowing, re-sowing, land preparation and weeding.

Local communities in Borena and South Omo employ different strategies to cope with the impacts of climate change-related problems. The strategies range from management of existing livestock and range and water resources to crop cultivation; use of wild leaves and fruits; and the introduction of camels. In this regard, the contribution of women is vital although they face an enormous burden that affects them in various ways.

The current institutional and policy responses focus more on emergency assistance, asset protection and livelihood enhancement, disaster risk reduction, and conflict management and resolution. Nonetheless, these responses hardly consider the gendered dimensions of climate change impacts in the area. The responses are designed to address immediate problems that pastoral communities face due to climate change instead of dealing with the underlying factors that aggravate long term adverse impacts. Generally, future interventions need to consider raising awareness in the areas of population planning, girls' education, women empowerment, reproductive health, and the impacts of harmful traditional practices which complicate the impacts of climate change.

References

Aklilu Amsalu and Alebachew Adem. 2009. Assessment of climate change-induced hazards, impacts and responses in the southern lowlands of Ethiopia. *FSS Research Report* No. 4. Addis Ababa: Forum for Social Studies.

Arndt, C.; S. Robinson, and D. Willenbockel. 2011, Ethiopia's growth prospects in a changing climate: A stochastic general equilibrium approach. *Global Environmental Change* 21,701-710.

Boserup, E. 1970. *Women's role in economic development.* New York, NY, St. Martin's Press.

Bouch, A.M. and Yared Mamo. 2008. Indigenous conflict management and resolution mechanisms on rangelands in Somali Regional State, Ethiopia. *Nomadic Peoples*, 12 (1): 109-121.

Brody, A., J. Demetriades and E. Esplen 2008. *Gender and climate change: Mapping the linkages, a scoping study on knowledge and gaps.* BRIDGE, Institute of Development Studies (IDS), UK.

Conway, D. and E. Schipper. 2011. Adaptation to climate change in Africa: Challenges and opportunities identified from Ethiopia. *Global Environmental Change* 21: 227–237.

Central Statistics Authority (CSA) 2008. Summary and statistical report of the 2007 population and housing census. Addis Ababa.

Dankelman, Irene. 2002. Climate change: Learning from gender analysis and women's experiences of organizing for sustainable development. *Gender and Development* 10, 21- 29.

Denton, Fatma. 2002. Climate change vulnerability, impacts, and adaptation: Why does gender matter? *Gender and Development* 10, 10-20.

Girum Getachew, Degefa Tolossa and Getachew Gebru. 2008. Risk perception and coping strategies among the Karrayu pastoralists of upper Awash Valley, Central Ethiopia. *Nomadic Peoples* 12, 93-107.

Hesse, C. and L. Cotula. 2006. Climate change and pastoralists: Investing in people to respond to adversity. *Sustainable development opinion.* London: IIED.

Horn of Africa Regional Environmental Center (HoA-REC) and GTZ. 2010. *Failing seasons, ailing societies: Climate change and the meaning of adaptation in Ethiopia*. Addis Ababa: Horn of Africa Regional Environmental Center.

Huq, S., H. Reid, and L. A. Murray. 2006. Climate change and development links. *Gatekeepers series* 123. London: International Institute for Environment and Development.

Intergovernmental Panel on Climate Change (IPCC). 2007. *Climate change 2007: The fourth assessment report of the Intergovernmental Panel on Climate Change.* Cambridge: Cambridge University Press.

Lambrou, Y. and G. Piana. 2006. *Gender: The missing component of the response to climate change.* Rome: Food and Agriculture Organization (FAO). Cited in Alyson Brody, Jestina Demetriades and Emily Esplen.

——.2008. *Gender and climate change: Mapping the linkages. A scoping study on knowledge and gaps.* Institute of Development Studies, UK: BRIDGE.

Nelson, V., K.; Meadows, T.; Cannon, T. Morton and A. Martin. 2002. Uncertain predictions, invisible impacts, and the need to mainstream gender in climate change adaptations. *Gender and Development,* Vol. 10, 51-59.

Oates, Naomi.; D. Conway, and R. Calow. 2011. *The 'mainstreaming' approach to climate change: insights from Ethiopia's water sector.* London: Overseas Development Institute.

Organisation for Economic Co-operation and Development (OECD). 2008. OECD environmental outlook to 2030. Available from: www.oecd.org/publishing/corrigenda. (accessed on 20th October, 2011)

Oesterle, Matthias. 2008. From cattle to goats: The transformation of East Africa Pokot pastoralism in Kenya. *Nomadic Peoples* 12, 81-91.

Röhr, Ulrich. 2007. Gender, climate change and adaptation: Introduction to the gender dimensions. Background paper prepared for the Both ENDS BRIEFING PAPER. Adapting to Climate Change: How Local Experiences Can Shape the Debate. Genanet – Focal Point Gender, Environment, Sustainability. Amsterdam.

Smucker, T. and B. Wisner. 2008. Changing household responses to drought in Tharaka, Kenya: vulnerability, persistence and challenge. doi:10.1111/j. 190 0361-3666.2007.01035.

Stige, L. C. *et al.* 2006. The effect of climate variation on agro-pastoral production in Africa, *PNAS* 103, 3049–3053.

Tiki, W.; G. Oba, and T., Tvedt. 2011. Human stewardship or ruining cultural landscapes of the ancient Tula wells, southern Ethiopia. *The Geographical Journal* 177 (1): 62-78.

United Nations Development Programme (UNDP)-GEF. 2010. Millennium Development Goals and climate change adaptation, the contribution [of] UNDP-GEF adaptation initiatives towards Millennium Development Goal 3, Issue No. 2, *Promote gender equality and empower women through climate change adaptation.*

United Nations Framework Convention on Climate Change (UNFCCC). 2007. *Climate change: Impacts, vulnerabilities and adaptation on developing countries.* Bonn: United Nations Framework Convention on Climate Change.

Ziervogel, G. and P. Ericksen. 2010. Adapting to climate change to sustain food security. *WIREs Climate Change* 1, 525-540.

Chapter 3

Challenged Livelihoods as a Result of Water Scarcity among Maasai Women Pastoralists in Kajiado County, Kenya

*Beth Njiru**

Introduction

Pastoralists are people who depend primarily on livestock as a livelihood system. They inhabit those parts of the world where the potential for crop cultivation is limited due to lack of rainfall and extreme temperatures (Rass 2006). Pastoralism offers a viable production system that enables huge arid and semi-arid areas to be used productively (Huho, Ngaira and Ogindo 2011). It determines the well-being of millions of the world's poor and supports a wide range of globally valued services and products, including biodiversity and raw materials (UNCCD 2007). There are about 120 million pastoralists in the world, of which about 50 million live in sub-Saharan Africa, many of them roaming the dry sub-Saharan belt that stretches from Mauritania to Ethiopia (Rass 2006). Statistics by the FAO in (2005) indicate that there were about 235 million cattle, 472 million goats, 21 million pigs, and 1.3 billion poultry in Africa, valued at US dollars 65 billion in 2005.

Water scarcity has emerged as the greatest global threat to nomadic systems. It is an increasingly scarce resource requiring careful management, with estimated fresh water resources of 21 billion m^3 (650 m^3 per capita) per year (UNESCO 2003). High

*bettynjiru2002@yahoo.com

temperatures lead to climate change that includes numerous effects, such as changes in precipitation patterns that can produce floods and droughts (UNEP 2010). According to IPCC (1995), warmer temperature will lead to a more vigorous hydrological cycle and this will translate into prospects for more droughts and/or floods in other places. Global climate change is likely to affect everyone on earth to some degree, whether in form of social, psychological, economic or environmental change, or even a combination of these (Baker, Ehrhart and Stone 2008). Projections indicate that there is going to be an increase in droughts in sub- Saharan Africa leading to more water scarcity (Campbell, Dalrymple, and Craig 2009).

Kenya is classified as a water-scarce country (National Report on Kenya 2006). Its per capita water availability is one of the lowest in Africa, and the situation is expected to get worse due to population growth and climate change. Access to clean water is already a problem in many areas of the country (Kandji 2006). This is worsened by the fact that most water for human consumption and other uses is derived from rivers whose recharge depends on rainfall (Ocholla- Ayayo 1997; UNESCO 2003; Kandji 2006). Estimates indicate that Kenya needed approximately 3.4 billion m^3 of storage of water by the year 2010 to reliably supply water for domestic and livestock development (UNESCO 2003). Already, Kenya has witnessed an alarming upsurge in the incidence and severity of extreme climatic events caused by climate change. Floods, droughts and landslides have ravaged virtually all parts of Kenya (GoK 2005; Republic of Kenya 2009). However water stress impacts are differently distributed among different regions, gender, occupa*tions, level of knowledge and adaptation strategies (Kandji 2006). Most people who will be invariably affected are the poorest and the most vulnerable communities* particularly women (Kandji 2006; IPCC 2007). Based on the above background, there was a need to study how water scarcity has challenged livelihood systems of Maasai women pastoralists of Kenya.

Statement of the Problem

The role of Maasai women pastoralists in coping with water scarcity is not clearly discussed although several studies demonstrate that women are the worst affected by water scarcity. Water scarcity is shaping the livelihood system of the Maasai women pastoralists. Taken to the edge by the increasing water scarcity, Maasai women pastoralists have devised numerous water scarcity coping strategies. Diversifying and adapting their livelihoods and environmental management can help support the pastoralist system in the long term in the face of increasing challenges (UNCCD 2007). A convention on biological diversity states that there is a need to understand and expose gender differentiated practices, gendered knowledge acquisition and usage. Gender considerations are not solely women's issues; instead this outlook could yield advantages for the whole communities and benefit both sexes (UNEP 2010). Women pastoralists lack awareness and information regarding climate change and the most effective and sustainable ways of coping with water scarcity, this has rendered most of the strategies they use unsustainable, thus worsening their state of affairs. Other strategies used in response to water scarcity share a great challenge that may have serious implications on the Maasai women pastoralists' livelihood (traditions) as well as the environment. Thus, there was a need to understand how water scarcity has challenged the livelihood systems of the Maasai women pastoralists. Bearing in mind that the livelihood system is more than just a set of physic-economic pre-conditions for continued existence, but also encompasses psycho-social dimensions of experience of living, it takes holistic consideration of things that the poor might be vulnerable to assets and resources that help them thrive and survive (Hiremath 2007).

Objectives of the Study

The purpose of this study is to establish the role of water scarcity in the changing livelihoods of Maasai women pastoralists of Oldonyo-orok sub-location. It focuses on women pastoralists, their changing livelihood systems and the coping mechanisms they have developed in their struggle to survive in the midst of diminishing water resources. The study's specific objectives were to:

- assess how water scarcity has challenged livelihood systems of the Maasai women pastoralists;
- find out the strategies used to cope with water scarcity by Maasai women pastoralists;

- explore the effects of the changed livelihood systems on resource use; and
- investigate the challenges women pastoralists are encountering in an attempt to cope with water scarcity and the challenged livelihoods.

Study Significance

According to the UNEP (2010) various studies have demonstrated that projects integrating gender dimensions generate superior results. This is because the role of women in development cannot be under-rated. The results of this study may give an insight regarding Maasai women pastoralists and how they are responding to water scarcity. This could lead to effective management and planning for both men and women in the society.

This being a relatively new area of study, it is expected that the findings will contribute to knowledge in this field about issues facing Maasai pastoralists as a result of water scarcity as well as additional research in the area.

The findings of this paper can be used to develop programmes that can be used to raise awareness, communicate knowledge, engage stakeholders, and support decision-making regarding climate change and its effects on the changing livelihood systems among women pastoralists.

Literature Review and Conceptual Framework

Approximately 80 per cent of Kenyan land is arid or semi-arid (ASALs) and only 20 per cent is arable land. Arid or semi-arid lands of Kenya occupy the country's largest portion, estimated to be 569,250 km^2 out of 582,650 km^2 of the entire country. The rest (13,400 km^2) is occupied by water (GoK 2005). Most of the land area is characterised by water scarcity (GoK 2005; Kandji 2006; Campbell, Dalrymple, and Craig 2009) – one of the hurdles preventing Kenya from achieving the Millennium Development Goals (MDGs), especially those related to poverty eradication and attainment of food security (Kandji 2006). Increasing temperatures and the high rate of dryness is already being felt in ASALs. Statistics indicate that temperatures in Kenya have risen by 1°C over the past 50 years, and warming is expected to accelerate, with temperatures rising nearly 3°C (Campbell, Dalrymple, and Craig 2009). The recent (2009) prolonged and severe drought is widely perceived to be symptomatic of the changing climate (*Ibid.*).

Livelihood Systems among Maasai Women Pastoralists

Livelihoods can be defined as the different ways in which people make a living and comprise the capabilities, assets, activities and strategies pursued by households and individuals for means of living (GoK 2005). Among pastoralists livelihood is derived mainly from livestock foraging on the natural vegetation, where opportunities for alternative livelihoods, including agriculture, are limited due to water scarcity (Berger 2003). Pastoralists generate most of their income from livestock rearing and, therefore, rely on the natural resources available to them like water and pasture (Campbell, Dalrymple, and Craig 2009; Witsenburg and Adano 2009). They share communal rangeland resources which are influenced by erratic rainfall that considerably varies between and within years (Witsenburg and Adano 2009). However, they are extremely susceptible to climatic hazards and hence, a risky undertaking for even the most experienced and knowledgeable herdsmen (Kareithi 2003).

The relationships between pastoralists, livestock, and the environment in the dry land of Africa are complex (Butt 2010). Pastoralist communities are marginalised on the basis of their geographical remoteness, their ethnicity, and their livelihood, which is still seen by many as an out-moded way of life that needs replacing with 'modern' livelihood systems (Oxfam 2008). The per capita livestock wealth of pastoral households has continued to decline over the years. As a result, pastoralists today rank high in terms of poverty levels and score poorly when assessed on other social welfare indicators, such as education levels, maternal health and nutritional status (Witsenburg and Adano 2009). Land adjudication and sub-division of group ranches has led to individual land tenure that has contributed to land sales to other people and, therefore, the opening up of the area to other communities from other parts of the country (CCAA 2009), thus pushing the local pastoralists to drier parts of the district (NCAPD 2005).

Pastoralists' Adaptation to Water Scarcity

Communities that live in risky environments such as drylands possess a rich repertoire of responses against climatic hazards (Kandji 2006). In the developing countries, pastoralists have shown throughout history a strong capacity for adapting to environmental changes (IPCC 2007). Over time, they have developed a highly efficient adaptation methods making Kenyan pastoral production one of the most efficient per capita (Kandji 2006; IUCN 2008). The harsh and unpredictable nature of the environment has led to the development of strong social obligations for use in times of drought, disease and conflict. Herd-sharing and splitting, the redistribution of surplus livestock within social networks, and the formation of alliances with neighbouring groups, all assist in protecting against destitution and in recovery from famine (Berger 2003). They also introduce new animal species to adapt to water stress. On various occasions pastoralists have been forced to change their livelihood completely, for example, shifting from pastoralism to other economic activities, such as small trade and handicraft (Kandji 2006).

While geography and climate largely explain Kenya's exposure to drought, the root cause of the country's vulnerability is its dependence on rain-fed livestock-keeping (Kandji 2006; UNEP & ICRAF 2006). Pastoralists are not entirely protected against, nor perfectly adapted to, current climate variability and extreme weather events such as drought (IPCC 2007). Women and men pastoralists face a series of challenges that hinder their way of life and stifle their ability to adapt to changes in their external environment (Oxfam 2008). The ability to adapt to environmental changes, to a large extent, depends on the level of economic development and the means required for adaptation (IASC 2009). Included in this is the high exposure to natural hazards, their direct dependence on climate sensitive resources such as plants, trees, animals, water and land, and their limited capacity to adapt and cope with climate change impacts (IASC 2009). Ochola-Ayayo (1997) indicates that in developing countries, the majority in agriculture and pastoralism depend on natural rain water for survival – a state where human beings are no better than animals; both simply wait for rain and do nothing to make water available. On the other hand, when there is too much of rain water (flood) they do nothing to store this water that can be used when there is no rain.

Gendered Roles among Pastoralists

Women constitute the largest percentage of the world's poorest people and the most affected by water scarcity (Brody, Demetriades and Esplen 2008). Despite this, it is paramount to note that gender roles affect economic, political, social and ecological opportunities and constraints faced by both men and women (UNEP 2010). Women remain overlooked in decision-making on the access to and the use of bio-diversity resources (*Ibid.*). Their contribution to development in developing countries is not quite appreciated (Mazonde 2001). In rural settings, women and men are often involved in multiple activities and different contractual arrangements simultaneously (Stonehill 2008). A closer examination of the division of tasks both within the household and the community reveals that there is more than merely the physical strength of the women in the division of such tasks; men use this division of labour to safeguard their cultural status (Mazonde 2001)

In almost all the communities in Kenya, fetching water is the responsibility of women, with the assistance of children, particularly female children (GoK 2005). Women and girls generally assume primary responsibility for collecting water for drinking, cooking, washing, hygiene and raising small livestock, while men use water for irrigation or livestock farming (Khosla and Pearl 2003; Fisher 2006). Society holds that such tasks do not contribute directly to the creation of jobs and do not generate income (Mazonde 2001). This does not mean that women do not play crucial roles in pastoral or other activities crucial to sustainable livelihoods and national economies. But the roles they play are generally less visible and attract less public recognition than the work men engage in (Brody, Demetriades, and Esplen 2008). What society fails to discern is that the many women who constantly perform these tasks make it possible for other people, particularly men, to do other duties. There is a real and strong positive relationship between the confinement of women to household chores and the gainful employment of their husbands (Mazonde 2001).

According to Hama Arba Diallo, Executive Secretary of the UNCCD, women play a pivotal role in the pastoralist way of life, assuming diverse responsibilities with regard to the livestock, the land and the household. In times of scarcity of natural resources, such as water while stress and hardship rise for everyone, it is women who are most burdened with the increased workload as they struggle to provide water for the family since it is a cultural responsibility for women to compensate (UNCCD 2007, 7). [Thus] natural disasters can provide women with a unique opportunity to challenge and change their gendered status in the society (WEDO 2008).

Effects of Water Scarcity on Maasai Women Pastoralist's Livelihood

New weather patterns and prolonged drought have emerged and pastoralists are now struggling with frequent water shortages which are threatening their impoverished livelihoods (Tiampati 2007). As traditional cattle herders, the Maasai men move out with the livestock leaving their homes for months at a time in search of pasture and water for their animals. In most cases, this means vulnerable women, children and the elderly are left behind to fend for themselves in the villages (Wamatsi 2009). Lennart Båge, President of IFAD, once said:

> "As soil fertility dwindles and crop and livestock productivity decline, men are forced to leave their communities in search of employment. Women are left to assume traditionally male responsibilities without the same access to community services; decision-making power; and financial, technical and other resources" (UNCCD 2007, 8).

According to WEDO (2008), as access to basic needs and natural resources like water is hampered, women's workload increases as they have to work harder to secure resources and livelihoods. In such situations, women have less time to earn income, get education or training, or participate in governing bodies while girls regularly drop out of school to help their mothers gather wood and water (GoK 2005; WEDO 2008; UNEP 2010).

Conceptual Framework

Figure 1 presents the conceptual framework used for analysing the relationship between water scarcity and changing livelihood systems among the Maasai women pastoralists. Linkages can be observed between water scarcity and response strategies, activity production (livestock, land water and pasture), causes of water scarcity, and changing livelihood systems.

Water scarcity is evidenced by drying rivers, erratic rainfall, prolonged drought and unpredictable rain. In addition, livelihood systems, activity production and strategies used to respond to water scarcity can, to a great extent, influence water availability. On the other hand, activity production will be influenced by the causes of water scarcity, coping strategies and livelihood systems.

Figure 1 also depicts that response to water scarcity (adaptation and coping strategies) will be influenced by activity production, causes of water scarcity, livelihood system and water availability. In other words, unpredictable rains will determine the kind of response. Similarly, livelihood systems of a Maasai woman pastoralist can be influenced by water scarcity, activity production and response strategies. To further illustrate this issue, availability of water determines the activity production, and whether or not a Maasai woman pastoralist will stick to pastoralism only, or diversify in or out of livestock production.

Correspondingly, a changed livelihood system will be influenced directly by water scarcity and livelihood systems. For instance, water availability will determine the source of income a woman pastoralist is likely to derive from livestock production; but with water scarcity she is likely to look for alternative sources of income.

Figure 1: **Relationship between water scarcity and changing livelihood systems among the Maasai women pastoralists**

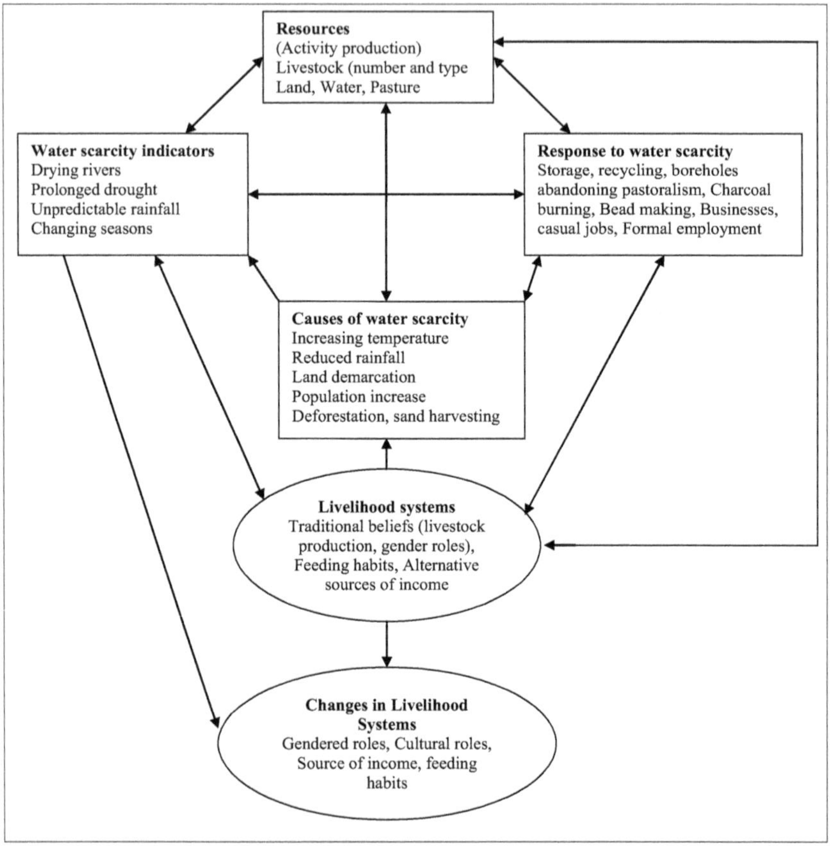

Source: Author's own construction

Methodology

This section describes the study site, target population, research design and approach, sampling procedures, and data collection and analysis procedures used in the study

The Study Area

The study was carried out in Oldonyo-orok sub-location of Namanga location, Kajiado County. Kajiado is an administrative district in the Rift Valley Province of Kenya located in the southern part of the province. According to Kenyan population and housing census of 2009, Kajiado County has a total population of 549,816 people (Kenya National Bureau of Statistics 2009). Oldonyo-orok sub-location has a population of 3,016 persons, comprising a total of 617 households with 1,538 females and 1,478 males, and covers an area of approximately 294.9 km^2 with a density of 10 (KNBS 2009). The sub-location has 12 villages, namely Enkidongoi, Engaboli, Intiatongoi, Enoosampurrumpur, Oloisuzua, Lositeti, Karero, Lenchani, Inkiwanjani, Parmatuni, Enkaroni and Olmani. This sub-location was selected on this basis:

Firstly, it is dominated by the Maasai who are largely pastoralists.

Secondly, given the diversity that exists in Kajiado County, with different groups of people ranging from urbanites to rural-based people, this livelihood zone was chosen as a unit of data collection because people share the same pattern of livelihood, that is, they are all pastoralists; they keep the same type of livestock; and they have the same way of living.

Study Sample and Data Collection Methods

A triangulated design, combining both quantitative and qualitative methods, was applied to this study. This design was used in the study to increase the credibility and validity of the result and to help in validation of data through cross verification from more than two sources: interview schedules and focus group discussions (FGDs). The study utilised the household as the unit of analysis.

A proportion of 10 per cent (that is, 62 households) was selected from the total 617 households in the study site. Equal allocation stratified sampling was then used to sample equal respondents from the 12 villages in the sub-location, that is, 62/12=5.16=5 (rounded off) households in each village. Systematic sampling was then used to select the households in each of the villages. This was done with an assumption that each village had equal number of households, therefore:

Total households = 617.

Total household per village = 617/12 = 51.4 = 52

Sampling constant = 52/5 = 10.4 = 10

A random number between 1 and 10 was then selected using blind folding technique. The number selected was 2 which then became the starting number. This ensured that all households in the village had an equal chance of being included in the study sample. The cases selected, therefore, included: 2, 12, 22, 32, 42 and 52. These case numbers were then used in each of the villages to select the household. This was done with the help of village elders who are conversant with the locality. To ensure that the sample had equal representation of male and female respondents, simple random sampling was used to select villages where only females were interviewed and villages where only males were interviewed. After the names of the villages had been written on a piece of paper, the blind folding technique was used to pick six villages at random where only males and/or only females were interviewed. This means that equal numbers (31 males and 31 females) were interviewed for the study.

On the other hand, purposive sampling was used to select FGD participants. A total of two FGDs were conducted, one for male participants and one for female participants. The group for males comprised 8 participants while that of females comprised 12 participants. Two (2) participants, a male and a female, were sampled from each of the 12 villages. Purposive sampling was used because it enables the selection of the cases rich in information.

Data collected from interviews and FGDs were supplemented through observation/visual assessment. This enabled the researcher to collect a depth of information about observable diversification, adaptation to water scarcity and environmental implications in the study area. A digital camera was used to record observations.

Before the commencement of the study, a research permit was obtained from the Ministry of Education (MOE). The researcher, with the assistance of two research assistants who had been given preparatory training, interviewed the household heads personally (one-on-one) by moving from household to household in the period between 18th October and 3rd November, 2011. Focus group discussion participants were requested to meet at a central place (Olmani Village) where discussions were conducted.

Data Analysis Techniques

To analyse quantitative data descriptive statistics, especially frequencies and relative frequencies (percentages), means and medians, were used. The frequencies and percentages were generated from the data set using STATA software. To test for association, the chi square (λ^2) test was applied. The chi-square was generated from the data set in order to determine whether there existed a systematic association or statistical significance between water scarcity and changing livelihood systems. FGDs were audio-recorded, transcribed, translated and digitally transferred to Atlas-Ti for content analysis. The data analysed are presented using tables and figures to ensure clarity in discussions and conclusion of the study.

Study Limitations

Data collection consumed more time and money than planned:

- Firstly, there were many call-backs since it was hard to get some respondents in their homes during the day.
- Secondly, in some instances research assistants were forced to follow the respondents in the market places and watering points.
- Thirdly, transportation is very poor in the sub-location and this made research assistants walk for long distances.
- Data collection was a cumbersome task due to excessive heat during the day and wild animals looking for water.

Results and Discussion

Background Characteristics of the Respondents

Overall, 62 households were studied. In each household one person was interviewed taking into account gender balance. Tables 1 and 2 present the characteristics of the respondents.

Table 2 shows that those interviewed had varied education levels with nearly half (48.4 per cent) having no education. The majority (58.1 per cent) of those who had not attained any education were women.

There was no significant difference between the average ages of male and female interviewees ($p = 0.79$).

In addition, there was no significant difference in the mean years lived in the region for males and females ($p = 0.68$).

Finally, there was no significant difference in the mean years of pastoral experience between males and females ($p = 0.50$).

Table 1: **Distribution of respondents, by age and sex**

Age and Sex	Number	Percentage
Age		
Less than 20 years	1	1.6
21 to 30 years	5	8.1
31 to 40 years	18	29.0
41 to 50 years	21	33.9
51 to 60 years	8	12.9
Over 60 years	9	14.5
Total	62	100
Gender		
Male	31	50.0
Female	31	50.0
Total	62	100

Source: Author's field data (2011)

Table 2: **Demographic characteristics of respondents**

Characteristic	Number	Percentage
Education		
Primary	18	29.0
Secondary	9	14.5
College	2	3.2
Diploma	1	1.6
Graduate	2	3.2
Total	32	51.5
Respondents with no education (n=30)		
Male	12	40.0
Female	18	60.0
Total	30	100
	Mean	Standard Error
Age	45.74	1.55
Years lived in the region	44.19	1.68
Years of pastoral experience	36.43	2.27

Source: Author's field data (2011)

Water Scarcity: Livelihood Systems Changes among Maasai Women Pastoralists

Evidence of Water Scarcity in the Region

According to the respondents, there are various indicators of water scarcity in the area. As evident from Table 3, 85.5 per cent of the respondents reported that rain amount has been on the decline with only 12.9 per cent reporting rain increase.

The locals reported receiving little amount of rainfall every season. Of the interviewees, 98.6 per cent felt that rain seasons had changed, thereby making rain and rain patterns unpredictable. It is no longer easy to tell the rainy months from the dry ones. In addition, it is hard to tell whether a particular season is for long rains or short rains.

Temperature was reported to have increased, with 83.2 per cent of respondents indicating that there has been an increase in the intensity of temperatures.

Table 3: **Rainfall and temperature changes in the study area (N = 62)**

Temperature and rain changes		Frequency	Per cent
Rainfall changes	Increased	8	12.9
	Decreased	53	85.5
	No change	1	1.6
	Total	**62**	**100**
Seasonal changes	Changed	60	96.8
	Not changed	2	3.2
	Total	**62**	**100**
Rivers	Dried	60	96.8
	Not dried	2	3.2
	Total	**62**	**100**

Source: Author's field data (2011)

Frequent and recurrent droughts have led to drying up of permanent water sources such as the streams, springs and wells. Permanent rivers that have provided water for years, including rivers Namanga, Oldonyo-rok and Engaboli have dried up. Figure 2 depicts the dry Namanga River bed. Respondents associated water scarcity with the increasing temperatures and the reduced amount of rainfall.

Figure 2: **Dried up bed of the Namanga River**

Source: Photo taken by the author (2011)

The findings of the household surveys concurred with those from focus group discussions. FGD participants reported to have noted changes in the amount of rainfall, increases in temperature and seasonal changes. The following excerpts from some of the FGD participants are illustrative:

> Earlier, there was no drought in Maasailand; rainfall was in plenty; livestock were many; and you could not hear of livestock death due to lack of water and pasture. But these days, the grass has reduced. These days you cannot walk for long with your animals because it is very hot, unlike before when we could walk for many kilometres with the livestock. Before in Maasailand, we used to have two rain seasons in a year; short rains came from October to November; and long rains from March to June. But these days, due to deforestation in order to burn charcoal, this has caused a lot of desert (Male FGD participant).

> During the 1980s, this area was very good to live in. Things were good and rainfall was in plenty and we had no problems. Cows and goats were uncountable. Men could get drunk and wake up the following day still rich; there was nothing to worry about. These days getting even 5 litres of water is a serious problem because the streams and springs have all dried up; rain no longer comes and when it does it comes for two days and disappears for the next one year. What kind of life is this? We are suffering; our animals are dying. A cow cannot give birth three times; it gives birth two times and dies as a result of funny diseases due to lack of water. We are growing poorer every day (Female FGD participant).

Asked what they thought was the cause of water scarcity, respondents gave a mixed reaction. Some of them reported it was as a result of reduced rainfall, increasing temperatures, population increase, land demarcation, deforestation, sand scooping and charcoal burning. However, the majority (69.4 per cent) pointed out lack of rain or insufficient rain as the major cause of water scarcity.

Similar sentiments were echoed during FGDs as evident from the following passages:

> Earlier Maasai people were sparsely populated and one home was around 10 km from another. This sparsely populated land was good for grazing but because of land demarcation, we do not even have a kilometre distance between households. Population increase and land sub-division have caused a lot of problems in Maasailand. The little water available is not enough and the sources available are owned by individuals who are selfish and cannot share water (Male FGD participant).

> We no longer get enough water because the population has increased and there were few people before. From here to the other side of the hill, there were only three homesteads but currently people are everywhere (gesturing). Land has been demarcated and it is an offence to graze your livestock on other people's farm. This has made movement hard. The forests are now guarded by the government officers because people are cutting trees for charcoal. Our animals have to be confined in a particular area (Female FGD participant).

It is evident that water scarcity is a great challenge in the region, raising a concern among the locals whose livelihoods depend on rain-fed water. This finding coincides with the findings of Tiampati (2007) that new weather patterns and prolonged drought have emerged and pastoralists are now struggling with frequent water shortages which are threatening their impoverished livelihoods. The locals can no longer predict rainfall seasons nor can they differentiate the long and the short rain seasons. They compare this occurrence with years back when there was a clear sequence of weather patterns. This scenario has influenced the livelihood systems and Maasai women pastoralists who, according to Berger (2003), derive their livelihood mainly from livestock foraging on the natural vegetation but are now limited due to water scarcity. Water scarcity has led to the collapse of the established 'traditional' sources of income from livestock production. In a broader sense, the livelihoods system (traditional livestock production) has been severely affected. On the other hand, Bevege (2009) notes that during the 2009 drought, the Maasai pastoralists who once sold their cattle for as much as Kenya shillings 30,000 per cow, now sold a cow at 1,000 shillings to buy feeds for the rest of their starving herds and for the family. A study by Nori and Davies (2006) also authenticates the findings that pastoralists depend on livestock for their livelihood. Any trigger that may affect any of this affects the whole livelihood system of the pastoralists.

Other than climate change, the other issues that were pointed out to be the cause of unending water scarcity in the area include population growth, land degradation, lack of traditional communal land, sand scooping and deforestation among others.

Maasai pastoral land is no longer communal as it used to be where one could move from one place to another with their livestock.

Traditionally, in Kajiado, most people practised the traditional Maasai mode of life that encouraged semi-nomadism and most land was communally owned (NCAPD 2005). This confirms the findings of CCAA (2009) that land adjudication and sub-division of group ranches has led to individual land tenure that has contributed to land sales; the movement of these traditionally nomadic peoples is severely hindered by conflict over grazing lands and water, and land degradation, as climate change increases the stress on their herds.

There have been numerous changes among the livelihood of the Maasai women as well as the locality. It is evident from the above presentation that these changes have been brought about by water scarcity in the region. The livelihood system of the Maasai pastoralist is dependent on rain-fed water. This confirms the findings of Kandji (2006) and UNEP & ICRAF (2006) that while geography and climate largely explain Kenya's exposure to drought, the root cause of the country's vulnerability is its dependence on rain-fed livestock.

All the households interviewed reported to own livestock but the number differed from one household to another. Nearly one third (32 per cent) of the households interviewed reported some form of cultivation along the water sources. All the respondents who reported some form of cultivation, however, pointed out that cultivation was no longer working in the area due to water scarcity.

Asked to enumerate the role of a traditional Maasai woman, the respondents gave a list of numerous traditional duties of a traditional Maasai woman, such as: fetching water, collecting firewood, feeding the children, milking, constructing the house, cook for the family and grazing the animals in the household where a man was not available. These findings are consistent with those by the GoK (2005) which found that, in almost all the communities in Kenya, fetching water is the responsibility of women with assistance of children, particularly girls. This was confirmed by focus group discussion participants as shown below:

> A Maasai woman's work is enormous; fetching water, firewood, milking and building houses. Once we move, we cut grass and build houses while the men construct cowsheds. The other thing is that, women have to walk for long distances to look for water unlike before when one could dig a shallow well for water. You may use donkeys but not all of us have donkeys. You are instead forced to carry water on your back.

> At times, it takes the whole day because you have to walk several kilometres – around 10 - 14 kilometres (Female FGD participant).

It is evident that most of the pastoralists depend on livestock production as livelihood means. This confirms the finding of Mwaniki, *et al.* (2007) that in Kenya the ASALs constitute about 80 per cent of the country's total surface area and supports 25 per cent of the total human population, and 5 per cent of the entire livestock population in Kenya. In Kenya livestock accounts for 95 per cent of the family income and provides employment for 95 per cent of the pastoral population (Nori and Davis 2006).

Cattle ownership remains high in Oldonyo-orok with all (100 per cent) interviewed households reporting cattle ownership at the time of the survey. Besides livestock production, Maasai women pastoralists are also engaged in other activities such as farming. This has proven hard, however, since the streams and water sources where they do farming have dried up.

It is clear that, to some extent, the role of a Maasai woman is dependent on livestock production. Water scarcity has abridged livestock production, thus changing the historical livelihood system. Masaai women pastoralists are now engaged in duties that were not meant for women in Maasai traditions such as making and hawking ornaments as well as burning and selling charcoal.

WEDO (2008) notes that natural disasters can provide women with a unique opportunity to challenge and change their gendered status in the society. The decreasing water resource is leading to the changing livelihood systems among the women who are now seeking alternative ways of supporting their families.

is also reveals that Maasai women are increasingly embracing education for their children since they no longer rely on livestock. Traditionally a Maasai girl child was engaged a few weeks after birth and this is referred to as *esirata*, meaning that the parents expected several animals from that engagement. These days this cultural practice is diminishing because there are no livestock to pay for the engagement.

Secondly, this has changed since the majority of women are taking their female children to school. The following quote from one of the female focus group participants is supportive of this trend:

> Many changes are occurring in Maasailand. Firstly, Maasai people have embraced education which is changing the livelihood of women. Education is now highly valued unlike in the past when a Maasai valued cows more than anything else. Secondly, we have embraced religion which is changing many things such as dressing, feeding styles and other traditions like early engagement.

Traditionally, the Maasai used to feed on meat and blood from their animals. According to FGD participants, a woman was supposed to feed on special meal made of blood and meat, referred to as *osaroi,* after delivery. She was also supposed to take a special type of fat from the animals referred to as *eilata*. Water scarcity has changed their feeding habits since there are no animals from which blood and fat can be drawn.

Respondents were concerned that women are not able to feed their children as they used to. A small baby was supposed to feed on animal cream called *engorno*. This cream was made from special type of animal fat. This has completely changed since there are not enough animals and the ones that are around are not sufficiently strong to produce enough of these products.

Other changes that were mentioned include reduced births, increase in the number of single women and a decline in polygamous families. The following quote from a male FGD participant illustrates the changes that have occurred in feeding habits among Maasai pastoralists:

> Initially a Maasai person could feed on blood and meat from the cows. This has changed with the grade cows (exotic breeds) because you cannot draw blood from a grade cow. It has forced our women to look for alternative food for the family like potatoes, maize and beans. During pregnancy and after birth, a Maasai woman was required to take special meal prepared from animal blood and meat. This no longer works because the animals are not enough.
>
> A traditional Maasai woman was supposed to stay in the house for nine months after giving birth. Her work was to eat and breastfeed the child. This has changed since a woman has to go out and do other jobs to earn income. These days, a woman will give birth and after a few weeks she is out doing other businesses to feed her children because livestock is not there and the ones that are there are not enough.

From the abovementioned issues, it is true to say that the livelihood systems of Maasai women pastoralists has greatly been affected, ranging from feeding habits, traditional roles, gendered roles, cultural roles and to social roles. This has been caused by water scarcity. From the finding, some slow but positive changes are imminent. Women are now embracing education for their children more than before while some cultural practices are fading away such as child engagement.

Coping Strategies Used by Maasai Women Pastoralists

All the respondents reported that they would wish to revert back to their 'traditional' livelihood.

98 per cent said they wished to be engaged in strategies that are enabling them to adapt to and/or cope with water scarcity. The adaptation strategies mentioned included increasing the number of goats and reducing that of cows; shifting from rearing traditional cows to grade cows; concentrating the animals at the watering areas; reducing the number of livestock; sinking boreholes, and water storage.

Other coping strategies include bead making, selling milk, producing and selling charcoal, sand harvesting, selling firewood, selling land, and hawking items in the town of Namanga. Figure 3 illustrates sacks of charcoal ready for sale while Figure 4 illustrates a Maasai girl child taking charcoal to the market using donkeys.

Figure 3: **Sacks of charcoal for Sale**

Figure 4: **A Maasai girl taking charcoal to the market**

Source: Photos taken by the author (2011)

Participants unanimously reported that water scarcity had changed the livelihood system of a Maasai woman. The patterns of livelihoods activities observed in the study area showed that there had occurred diversification with many pastoralists having digressed from the "traditional" pastoralist way of life. Asked whether they were engaged in other income-generating activities apart from pastoralism, 90.3 per cent of female respondents indicated that they had diversified their livelihood systems, compared to 45.2 per cent of men respondents who reported being engaged in other income-generating activities (see Figure 5). To illustrate this point, other activities that formed an important part of pastoralists' livelihoods included making of beads which are hawked in the town of Namanga and other towns and selling charcoal and firewood, among others. A male focus group participant had the following to say about such activities:

> Both men and women have been forced to engage in varied businesses as a result of drought. Traditionally, men were supposed to take care of the livestock and look for water sources. Women were supposed to stay in the house looking after and feeding the children, milking animals, fetching water and firewood. Then animals were many and catered for all your needs and there was adequate rainfall. These days, men and women are engaged in small businesses so as to get income.

Figure 5: **Diversification of livelihood systems among the Maasai, by sex**

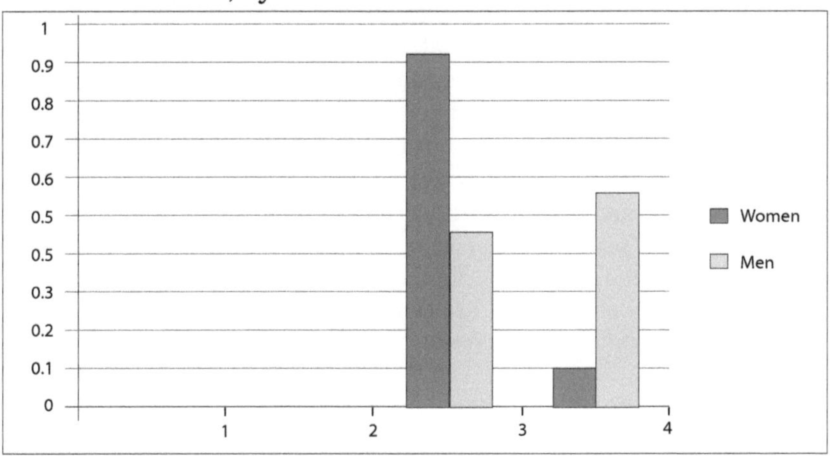

Source: Author's field data (2011)

It cannot be denied that diversification and adapting to water scarcity is the only option for women pastoralists. The above presentation indicates clearly that pastoralists have adopted various strategies in order to survive. According to Huho, Ngaira and Ogindo (2011), the inhabitants of the ASALs live in harsh and very demanding environments and often exist on the brink of survival. As of necessity, they have evolved a variety of drought coping and adaptive strategies which enable them to survive drought periods. In the same way, the locals of Oldonyo-orok have developed survival mechanisms to make it possible for them to survive the harsh conditions in their locality. These communities have no choice but to continue to use the resources at their disposal to adapt and survive (Jon and Rachel 2009). The study has also revealed that women are more likely to diversify their livelihood options compared with their male counterparts. This confirms the findings of UNCCD (2007) that, given the chance, women have resourcefulness and adaptability in abundance, and they are highly capable of finding alternative sources of income, particularly during periods of intense drought in the drylands, and of acquiring the necessary skills to do so. This is true for the Maasai women pastoralists of Oldonyo-orok.

Effects of the Changing Livelihood Systems on Resource Use

Droughts in Oldonyo-orok have led to numerous problems among the locals. According to the respondents, poverty and death of animals are the greatest effects of water scarcity in the area. Asked to enumerate the effects of water scarcity on pastoralism, 82.3 per cent of the respondents reported that water scarcity had led to animal death. Other effects that were enumerated included migration (41.9 per cent), human conflict (32.3 per cent), water-borne diseases (22.6 per cent) and human-wildlife conflict (17.7 per cent) (see Table 4). The many effects posed by water scarcity were also supported by data from focus group discussion as evinced in the following excerpt:

> Problems of water scarcity are numerous; goats and cows are dying. Animals stay for days at the borehole area without water. The weak people are dying due to lack of water; you may have food but you cannot cook because there is no water. This is because the food that is being eaten by Maasai people today requires water unlike before when a Maasai could only rely on gourd for milk and blood and roasted meat. There were

no utensils to be washed and no food to be cooked with water. Water scarcity is making people concentrate at a particular area where there is a wetland. One may wish to move but you cannot because there is no water. This is degrading the areas where animals are concentrated. Water scarcity is also causing conflict. During the drought people may even fight or kill each other; one wants his animals to take the little water available. This has also made people get up as early as midnight to take their animals to drink water before other pastoralists get up. We are supposed to follow schedules but this is causing clan conflicts because one clan has its own day to go to the watering point. It has led to clan boundaries (Male FGD participant).

Table 4: **Distribution of responses regarding effects of water scarcity on pastoralism (N = 62)**

Effects of water scarcity		**Frequency**	**Per cent**
Animal death	Animal death No death	51 11	82.3 17.7
	Total	**62**	**100**
Migration	Migration No migration	26 36	41.9 58.1
	Total	**62**	**100**
Human conflict	Conflict No conflict	20 42	32.3 67.7
	Total	**62**	**100**
Water borne diseases	Water borne diseases No diseases	14 48	22.6 77.4
	Total	**62**	**100**
Human wildlife conflict	Human-wildlife conflict No conflict	11 51	17.7 82.3
	Total	**62**	**100**

Source: Author's field data (2011)

Asked whether they share water with livestock, 58.1 per cent reported that they did while 38.7 per cent responded in the negative.

Concerning whether changed livelihood has any effect on the use of water, pasture, and on the number and type of livestock, all the respondents interviewed agreed with the statements. On the effects of changed livelihood on water resource use, the interviewees reported that in many cases women were forced to buy water since they did not have enough time to look for water. This is because they had to engage in other income-generating activities such as hawking ornaments in town. Women are also buying storage tanks to store water in case it rained, something they were not doing before.

Regarding the pasture use, some of the respondents stated that they were buying pasture from as far as Tanzania for their livestock since their movement had been curtailed by land demarcation. Others stated that they were growing pasture on their farms to ensure continuous availability – something they had never done before.

The number of livestock is reducing in various households and most of the pastoralists are now engaged in activities for alternative income such as sand scooping and the burning and selling of charcoal. Traditionally, land was communal property but these days land has been demarcated and many pastoralists are selling their land. This has forced them to reduce the number of livestock. Women are now selling milk unlike in the past. This has led to the change of the type of animals, with a number of pastoralists now resorting to grade cows. More goats are now taking the place of livestock. This is evident from the words of a female focus group participant, who stated:

> "Before, we could not sell milk in this region both from cattle and goats. But these days milk is making a lot of money. Everybody is asking for goat milk; that's why it is good to keep many goats".

It is apparent that many pastoralists rely for all or part of their livelihood on common property or publicly-owned resources, such as forests, water resources and common grazing land. Availability of these resources, in turn, is dependent on water. Research indicates that pastoralists directly depend on climate-sensitive resources, such as plants, trees, animals, water and land, and their limited capacity to adapt and cope with climate change impacts (IASC 2009). Water scarcity in the region then has shaped the use of resources. Perhaps the greatest obstacle to resource use is where people have to share water with the animals. All of these are the effects of inadequate water in the region. A study by GoK (2005) reveals that in the ASALs, residents are forced to share the same water sources with livestock due to water scarcity.

Traditionally, many livestock signified wealth among the Maasai. This has been altered by the changing livelihood caused by water scarcity. Pastoralists have been forced to reduce the number of their livestock and modify the type of animals they keep.

The positive side of these findings is that women pastoralists are now storing water after rains – an undertaking that can be beneficial if adopted by all.

Coping with Water Scarcity and Changing Livelihood Challenges

One of the greatest challenges as a result of changed livelihood systems is poverty. Insights from qualitative data reveal that social instability is on the rise. Male FG participants blamed it all on women who have abandoned their "traditional" roles and are doing other businesses.

Diseases, particularly HIV/AIDS, have also been reported to have increased. All has been left to the women and, therefore, a woman has no choice but to look for other sources of income.

Women FG participants expressed humiliation and anger over men being unable to maintain their role as bread-winners, who, they said, are engaged in alcohol consumption instead. The women tended to blame men for their plight. As one female FDG participant aptly put it:

> These days men have abandoned their role in the household. It is women who are supposed to feed and educate the children. Most men have sold their land and spent the money on the prostitutes in town, and come home with diseases. Others are taking beer from morning to evening and this is because they have no animals to graze.

Asked whether there were challenges they were facing when trying to regain their livelihood system, all the respondents pointed out numerous challenges. They, for instance, reported that they lacked income which was a hindrance to survival strategies. In addition, they lacked tools to sink boreholes; had no expertise regarding some adaptation strategies (for example, keeping grade cows); lacked storage tanks and iron sheet roofs from which they could harvest rainwater.

Furthermore, some of the products, for example ornaments made by women, lack ready markets. Focus group discussions also indicated that women have limited time to do their household chores since most of the time is spent away from home.

Respondents acknowledged the pressure their activities are placing on the local environment, but they have no alternative. Those interviewed reported that some of the survival strategies are causing environmental problems directly or indirectly. For example, concentrating animals in one area is leading to soil degradation while sand harvesting is exposing water bends increasing evaporation and reducing the water available. Conversely, the run off from the exposed bends is causing soil erosion and thereof soil degradation.

The indirect problems that were reported are deforestation in the area with most of the people burning charcoal. This is leading to reduced rain, hence, water scarcity. Figure 6 shows an area of the ground from which sand has been scooped. It should be underlined that the change of the livelihoods system is likely to have dramatic and potentially negative impacts on both pastoralists themselves and their social and natural environment. Even though pastoralists are trying to adapt as well as look for alternative livelihoods, there are obstacles that are preventing their success. Many of the strategies are not working well and are worsening the already fragile situation.

Women are stepping outside of their household responsibilities to earn a living and help put food on the table. Numerous barriers exist due to the newly adopted duties that have left Maasai pastoralist with little time to carry out other duties. In such situations, women have less time to earn income, get education or training, or participate in governing bodies while girls regularly drop out of school to help their mothers to gather wood and water (GoK 2005; WEDO 2008; UNEP 2010). However, this has increased the household responsibility of women who spend most of their time selling and hawking ornaments, burning charcoal and looking for water. In times of scarcity of natural resources, such as water while stress and hardship rise for everyone, it is women who are most burdened with the increased workload as they struggle to provide water for the family since it is a cultural responsibility for women(UNCCD 2007, 7; WEDO 2008). The risks and vulnerabilities of the poor who live in insecure places and need to build their resilience to cope with climatic fluctuations are among the most important challenges in adapting to increasing water scarcity (FAO 2007).

According to the pastoralists, the solution to the problems they were facing when trying to cope with water scarcity and changing livelihood challenges lay with the government providing funds and tools for digging boreholes, ban cutting of trees and charcoal burning. People should also be trained on various adaptation strategies such as keeping grade cows and planting trees. Women initiatives such as bead-making and selling goat milk should also be supported.

Conclusion and Policy Recommendations

Conclusion

The main objective of this study was to explore and determine how water scarcity has challenged the livelihood systems of the Maasai women pastoralists. The study further sought to find out the effects of challenged livelihood systems on resource use in the study area, survival strategies used by the pastoralists, and the challenges they are encountering in response to water scarcity.

It is evident that most of the respondents, as pastoralists, depend solely on livestock. Thus, cattle ownership remains high in Oldonyo-orok with all households reporting cattle ownership at the time of the survey. The increasing water scarcity in Maasailand is indeed changing the livelihood system of the Maasai women pastoralists. This has led to the instability of the traditional livelihood systems of the poor pastoralists who have always depended on livestock production.

The findings reveal that water scarcity is caused by several driving forces. However, the direct causes that were mentioned include declining rainfall, increasing temperatures and seasonal changes.

The indirect causes can be presumed to be population growth, land demarcation, cutting down of trees, charcoal burning and sand harvesting. For instance, population has increased and so has the number of livestock. This has led to over-use of the available water resource in the area. Land demarcation has restricted nomadism in the area placing some water resources such as springs in the hands of a few people who sell the water.

The study also shows that some of the feeding systems of pastoralists have been affected and this is again associated with water scarcity. Specifically, feeding habits have changed and women and other family members feed on foods such as maize and beans as a replacement for traditional foods (blood, milk and meat). The study reveals that there are some slow but positive changes among the Maasai women's livelihoods. For instance, some cultural practices such as female children engagements are slowly fading and women are now embracing education more than before.

It is evident from the preceding discussion that Maasai women pastoralists are now taking over men as household heads. Overall, levels of livelihood diversification are significantly higher among women compared with men pastoralists. More women are deviating from the traditional norm and are now engaged in other businesses. The findings reveal that diversification away from "traditional" pastoralism has been triggered by water scarcity and has been adopted more by women. Two types of diversification can be deduced from the findings. These are:

> 'in' or 'within' livestock diversification where women pastoralists combine livestock production and other income-generating activities, such as bead making and selling milk, and
>
> 'out' or 'without' livestock diversification where women pastoralists have completely abandoned pastoralism and are now engaged in other income-generating activities, which mostly is done by those who acquire formal jobs.

The former is the most common form of diversification.

As revealed in the study findings, a Maasai woman is the one responsible for obtaining water and firewood, constructing the house, cooking for the family, as well as looking after the young ones; thus her vulnerability to water scarcity is very extensive. Any change in water availability can adversely affect the women given that most of their roles are water reliant.

Water scarcity is increasingly influencing the resource use. Water availability determines the number of livestock, the type of livestock and livelihood systems. The changing livelihood system has greatly influenced the traditional use of natural resources (water, pasture/land and livestock system). Interview with the Maasai pastoralists has revealed that the majority of them are now abandoning livestock production and this is influencing resource use in the area. With adequate water, animals are able to sustain pastoral communities' livelihood systems.

The reduction of the number of livestock and changing from cows to goats is a clear revelation that water scarcity has greatly affected resource use.

Traditionally, the Maasai livelihood system has been maintained by the communal nature of land tenure. The switch from the communal land tenure system has changed the Maasai rangelands which are now demarcated and increasingly privatised. Land demarcation has also contributed to the reduced livestock movement. Some pastoralists are selling their land to venture into other sources of income. Most of the respondents felt that land demarcation has curtailed their nomadic way of life.

The use of pasture has changed with pastoralist's now growing pasture on their farms in addition to buying pasture from areas with plenty.

Boreholes are increasing in the area with most of the people now using borehole water, unlike before when the source of water was springs, rivers and wells.

Branching away from pastoralism is indeed becoming common among the women pastoralists of Oldonyo-orok but not without effects on their livelihoods and the environment.

Although women pastoralists have diversified their livelihoods, they are faced with innumerable problems. They have limited opportunities since some of their activities require substantial amount of income which many cannot afford. Let us cite some examples:

- A grade cow is very expensive and maintaining it is even more expensive.

- Although pastoralists have tried to cope with water scarcity for many generations their adaptability has not enabled them to cope with the growing challenge of water scarcity which seems to be worsening by day.
- They lack relevant information on environmental changes and adaptation to water scarcity.
- It is also worth noting that even though women may be engaged in other income-generating activities besides their household chores, the demanding responsibility of running household activities remains largely with them. This is restraining productivity in other income-generating activities that are used as a way of coping with water scarcity.

The study reveals that some of the strategies used to cope with water scarcity are causing environmental problems. The participants know this too well but they lack alternatives. For instance, cutting down of trees has reduced the natural capacity of rain attraction while sand harvesting has reduced the capacity of the soils to hold rain water as well as causing soil degradation. Interestingly, the study reveals that most of the respondents blame it all on the government failure to help. They point out that the government holds the solutions to problems they are experiencing. For example, it was indicated during the FGDs that the government should provide funds and tools for sinking boreholes. The government should train pastoralists more in water management and tree planting. They also pointed out that there is a need to ban charcoal burning and cutting trees.

Policy Recommendations

Altered livelihood system for pastoralists is faced with numerous challenges and, therefore, requires adaptation and coping information and involvement in the planning and management of water resources. Investment in information and capacity-building to ensure awareness so as to increase adaptive capacity is required. This will increase capability of pastoralists, strengthen their asset base, obtain higher returns on their assets and attain more secure livelihoods.

Women pastoralists make an important contribution towards the economy, therefore, with the right support, they can cope with water scarcity and the challenges they are facing in an attempt to change as well as take up the challenged livelihood systems. There is a need to explore more on the diversification strategies that women pastoralists are using. Information on such issues can be crucial to take effective, coping and adaptation as well as poverty reduction measures. Gender considerations are not solely women's issues; instead, this outlook could yield advantages for the whole communities and benefit both sexes (UNEP 2010).

There is a need to maintain and enhance the quality of land resources through provision of assets and livelihood means such as piped water and boreholes.

Additionally, providing poor pastoralists with finances is an important way of breaking the poverty cycle.

Income is a major determinant of adaptive capacity. Therefore, the government needs to assist women pastoralists who are vulnerable to water scarcity in meeting the costs of adaptation to those adverse effects.

Right skills determine how successful adaptation and coping strategies are. There is a need to build capacity and transfer technologies in the regions so as to enhance adaptation. Improving skills at the community level is one of the pre-requisites of adaptation. This can be done, for instance, by improving the efficiency of existing traditional livestock production system instead of creating new ones and shifting away from traditional approaches. It is of paramount importance to improve existing traditional approaches to livestock production.

There is a need to establish partnerships with various organisations and government departments such as the Ministry of Trade so as to avail markets for the products made by the Maasai women pastoralists.

Acknowledgment

I acknowledge the support given by the Administration of Namanga location, more specifically the Oldony-orok sub-location. Also many thanks go to my research assistants.

Finally, I would like to send my sincere thanks to all communities and village leaders who contributed to this study.

References

Baker, J.; Ehrhart, C. and Stone, D. 2008. Hotspots -prediction and action. In *Forced Migration Review. Climate Change and Displacement,* Issue 31, Edited by Marion Couldrey & Maurice Herson ,44-45 October 2008. Oxford: University of Oxford.

Berger, Rachel. 2003. Conflict over natural resources among pastoralists in northern Kenya: A look at recent initiatives in conflict resolution. *Journal of International Development,* 15, 245–257 (2003). Rugby: Published online inWiley InterScience (www.interscience.wiley.com). DOI: 10.1002/jid.985

Bevege, A. 2009. Drought in Kenya ravages Maasai cattle. In *Indigenous Portal.* Available at: http://www.indigenousportal.com/Climate-Change/Drought-in-Kenyaravages- Maasai-cattle.htm (accessed on October, 2011)

Brody, A.; Demetriades, J. and Esplen, E. 2008. *Gender and climate change: Mapping the linkages a scoping study on knowledge and gaps BRIDGE.* Oxford: Institute of Development Studies (IDS).

Butt, B. 2010. Pastoral resource access and utilization: Quantifying the spatial and temporal relationships between livestock mobility, density and biomass availability in Southern Kenya. *Journal of Land Degradation & Development.* Volume 21, no. 6, pp.520–539.

Campbell, Ivan; Dalrymple, Sarah and Craig, Rob. 2009. *Climate change and conflict.: Lessons from community conservancies in northern Kenya.* Oxford: Conservation Development Centre, International Institute for Sustainable Development and Safer world.

Climate Change Adaptation in Africa Program (CCAA). 2009. *Enhancing adaptation to climate change among pastoralists in Northern Kenya.* Nairobi: CCAA.

Fisher, J. 2006. *For her it's the big issue: Putting women at the centre of water supply, sanitation and hygiene.* Geneva: Water Supply and Sanitation Collaborative Council.

Food and Agriculture Organization (FAO). 2007. *Adaptation to climate change in agriculture, forestry and fisheries: Perspective, framework and priorities.* Rome: FAO.

----.2005b. *Global Livestock Production and Health Atlas.* Rome: FAO

Government of Kenya (GoK). 2005. Kajiado District strategic plan 2005-2010 for implementation of the National Population Policy for Sustainable Development Nairobi: NCAPD.

Hiremath, B. N. 2007. The changing faces of rural livelihoods. In: The National Civil Society Conference of What it takes to Eradicate Poverty, December 4 – 6, 2007. New Delhi, INDIA.

Huho, M. J.; Ngaira, J.K.W. and Ogindo, H.O. 2011. Living with drought: The case of the Maasai pastoralists of northern Kenya. In *Educational Research*, vol. 2(1), pp. 779-789, *International Research Journals*. FAO.

International Association for Scientific Knowledge (IASC). 2009. Climate change, food insecurity and hunger. Technical Paper of the IASC Task Force on Climate Change. Submitted by WFP, FAO, IFRC and OXFAM, as well as WVI, CARE, CARITAS, WHO and Save the Children, October 2009.

Intergovernmental Panel on Climate Change (IPCC). 1995. IPCC Second Assessment Climate Change 1995. A Report of the Intergovernmental Panel on Climate Change. The Science of Climate Change WMO, contribution of Working Group. Available at: <www.ipcc.ch/pdf/climate-changes-1995/ipcc.../2nd-assessment-en.pd>

——.2007. Climate change: Impacts, adaptation and vulnerability: IPCC Working Group II Report, Chapter 19, IPCC. Available at: <http://www.ipcc.ch/pdf/assessment-report/ar4/wg2/ar4-wg2-chapter19.pdf>

International Union for Conservation of Nature (IUCN). 2008. *A global perspective on the total economic value of pastoralism: Global synthesis report based on six country valuations*. Nairobi: IUCN.

Jon, E. and Rachel, B. 2009. *Community-based adaptation to climate change: Participatory learning and action 60*. IIED, December 2009. Oxford: Russell Press.

Kandji, Serigne T. 2006. Drought in Kenya: Climatic, economic and socio-political factors. In *The new standpoints*, World Agro-Forestry Centre, pp. 17-19.

Kareithi, J. N. 2003. Change and continuity in traditional pastoralism in Kenya: Environmental stress and food security. In The papers for population issues health, political economy, genders and environment, *Occasional Publication Series*, vol. V, pp. 245-253. Nairobi: PSRI.

Kenya National Bureau of Statistics (KNBS). 2009. *Kenya population and housing census (KPHC). 2009: Population distribution by age sex, and administrative units*. Nairobi: Kenya National Bureau of Statistics.

Khosla, P. and Pearl, R. 2003. *Untapped connections: Gender, water and poverty*. New York: WEDO.

Mazonde, N. I. 2001. *Gender in African society. Population, health and development in Africa*: Anthropological perspective.Nairobi: Population Studies and Research Institute.

Ministry of Environment and Mineral resources. 2009. *Understanding and responding to climate change*. Nairobi: Government press.

Mwaniki, Terry W.; Mbuchi, Peter M.; Leleruk, Marl and Mwei, Fred. (eds.). 2007. *Peace building and conflict management: Joint case study of North Rift region of Kenya*. Nairobi: Pauline Publishers.

National Coordinating Agency for Population and Development (NCAPD). 2005. Kajiado District Strategic Plan 2005 – 2010: For Implementation of the National Population Policy for Sustainable Development. Nairobi: Government press.

National Report on Kenya. 2006. Agrarian reforms and rural development new challenges and options for revitalizing rural communities in Kenya. The International Conference on Agrarian Reforms and Rural Development. Port Alegre:FAO

Nori, Michele and Davis, Jonathan. 2006. *Change of wind or wind change: Climate change, adaptation and pastrolism*. Nairobi: IUCN.

Ochola-Ayayo, A.B.C. 1997. The implications of our creative diversity in global ethics. Culture and development for the 21st century. UNESCO. 24-25, February 1997.

Oxfam International. 2008. Survival of the fittest: Pastoralism and climate change in East Africa. Oxfam Briefing Paper, August 2008.Oxford: Oxfam International.

Rass, Nikola. 2006. Policies and strategies to address the vulnerability of pastoralists in sub-Saharan Africa. *PPLPI Working Paper* No. 37. Rome: Food and Agricultural Organization.

Stonehill, A. 2008. World water crisis. Z *Magazine*, 19June 2008.

Republic of Kenya. 2009. *National climate change response strategy.* Ministry of Environment and Mineral Resources. Nairobi: Government Press.

Tiampati, Michael. 2007. Climate change sounds death knell to pastoralism. 7th September 2007. Nairobi: Mainyoito Pastoralist Integrated Development Organization.

United Nations Environment Programme (UNEP). 2010. *Biodiversity is life.* Convention on Biological Diversity, 2010. International Year of Biodiversity. Canada: Natural Sciences Sector.

UNEP & ICRAF. 2006. *Climate change and variability in the Sahel region: Impacts and adaptation strategies in the agricultural sector.* Nairobi: United Nations Environment Programme and World Agro forestry Centre.

United Nations Educational, Scientific and Cultural Organization (UNESCO). 2003. *Water for people: Water for life.* Paris: UNESCO Publishing.

United Nations Convention to Combat Desertification (UNCCD). 2007. *Women pastoralists preserving traditional knowledge facing modern challenges.* Bonn: Secretariat of the United Nations Convention to Combat Desertification.

Wamatsi, N. 2009. Africa climate change threatens life and health of Maasai women. Women News Network – WNN. accessed on http://womennewsnetwork.net (accessed on October, 2010)

Women's Environment & Development Organization (WEDO). 2008. Communicating changing risks. In *The Climate Change and Displacement Issue, 31,* 56. Oxford: University of Oxford.

Witsenburg, Karen and Adano, Wario. 2009. *Of rain and raids: Violence livestock raiding in Northern Kenya.* Oxford : James Currey.

Chapter 4

Gendered Impacts and Adaptation Mechanisms to Climate Change among Afar Pastoralists in North Eastern Ethiopia

Mulubrhan Balehegn and Kelemework Tafere[*]

Abstract

The level of impact of climate change and adaptive capacity is different for different sections of the pastoral community. A combination of different techniques, including focused group discussions, individual interviews, case studies, and structured observations, were used to see if there is difference in the level of vulnerability and adaptive capacity between the Afar men and women. Results indicate that, gender inequality is inherent in the Afar customary tradition (*Adda*), and acts as a risk multiplying factor, increasing the vulnerability of women to the ills of drought. Afar women and men have varying opportunities and limitations towards adaptation to climate change, with women playing a significant role in household level adaptation. Activities alien to the Afar culture, such as labour migration, petty trading and individual wealth accumulation, have emerged due to climate change. It is concluded that a gendered approach that recognises the potentials, limitations and vulnerabilities of different demographic sects of the pastoralist communities is essential for a better adaptation in the future.

Keywords: *Climate Change Impact, Adaptation, Gender, Afar Pastoralism*

[*] mulurhan.gebremikael@student.umb.no, kelem40@yahoo.com

Introduction

Background and Problem Statement

With increasing recognition that climate change affects sustainability of economic and social developments worldwide (Denton 2002), the subject is currently high on the agenda of international policy-making, research and development. This chapter is an attempt to build on existing literature on climate change impacts and gendered adaptation in the context of Afar pastoralism.

There is growing evidence indicating that climate change will speed up weather extremes, leading to increased vulnerability of humans to devastating catastrophes (Hassen 2008; Songok*et al.* 2011). Severe climatic shift is particularly expected among communities in semi-arid land whose livelihood mainly depends on rainfall. This culminates in the loss of household assets and livelihood opportunities and incapacitates local coping mechanisms that are important to maintain food security and socio-economic development (Songok*et al.* 2011). There is now a tendency to find implicit or explicit linkages between adaptation and development-focused action, especially with the International Panel on Climate Change (IPCC)'s signal on the disproportionate vulnerability of developing countries to climate change and the lack of adaptive capacities thereof (IPCC 2007).

In Africa, climate change is particularly putting severe pressures on the dry land areas and the people occupying those areas. While it is likely that people across various locations will experience differential impact of climate change and the long term effects may be hard to forecast, many climate change models foresee mounting levels of temperature and a declining trend in rainfall in several dry land locations. Already, dry environments get drier with severe shortages of water, unpredictable and uneven rainfall patterns as well as extreme weather conditions manifested through extended and regular drought periods. These developments shake the already elusive balance on which the pastoral mode of production is grounded (Hese and Kotula 2006). In the East African region, impacts of climate change on pastoral livelihoods are further complicated by such factors as political and economic marginalisation, inappropriate development polices, and increasing resource competitions among multiple users which exacerbate their vulnerability to drought-induced shocks (Neely, Bunning, and Wilkes 2008).

Although climate change affects every human being on the globe, its impactsare highest on the most vulnerable and marginalised social groups who have little opportunity to prepare for them and participate in negotiations to prevent them. These marginalised groups, notably women constitute the bulk of the world's poorest, require meticulous adaptation strategies (IPCC 2007). The gender equity and aspects of climate change are likely to be multi-faceted. The magnitude and scope of these aspects can be analysed in terms of inequalities in gender and development; the gender impact of climate change in various parts of the world and across different agro-ecosystems; aspects of gender participation in climate negotiations; gendered impacts of mitigation activities, and gender-based differences in impact of adaptation and mitigation (Terry 2009). Gender norms and power inequalities that are pertinent to participation in decision-making politics, the division of labour, access to and control over resources, and knowledge and skills entrenched in the social structure are important factors in climate change because they affect adaptation capabilities of men and women to climate risks (Long and Long 1992). On the other hand, climate change is believed to have a compounding effect on existing gender-based inequalities necessary to access to basic resources such as income, technologies, education and information (Masika 2002).

Effective mitigation of climate change impacts presupposes gender sensitivity in consultations and decision-making with emphasis on specific needs of vulnerable men and women. Besides, the capacities of poor people in contributing their crucial insightful knowledge and experience should be adequately recognised. Women are likely to be at a disadvantage with regard to participation in decision-making processes at all levels of society (household, community, national and international). This could be due to the lack of confidence and/or resources to contribute. But equally important is the fact that whatever contribution women make it is often down-sized (*Ibid.*).

Unfortunately, incorporating the issues of climate change in pro-poor planning, and more importantly, the establishment of the linkage between climate change and gender has been given a marginal importance until recently (Lambrou and Piana 2006). Despite recommendations on establishing such a linkage (Wamukonya and Skutsch 2002), research dealing with gendered dimensions of climate change is still scanty (Brody, Justina, and Emily 2008; Dhanashri 2010). Just as gender is not sufficiently mainstreamed in many areas of development policy and practice, the potential impacts of climate change on gender relations have not been aggressively dealt with and remain invisible. The situation in Ethiopia is not an exception. There is still a long way to go with regard to mainstreaming of gender and climate change issues. For instance, according to Brody, Justina, and Emily (*Ibid.*), out of 15 African Adaptation Programme (AAP) countries, based on the degree to which gender aspects are incorporated in climate change adaptation, Ethiopia ranks only within the lowest three countries.

Objectives of the Study

The study is conducted with the following specific objectives in mind:

- To examine the overall views, perceptions and attitudes of the Afar community towards climate change;
- To examine drought impacts on livestock, the bio-physical environment and social relations in Afar;
- To study gender-based impacts of climate change among Afar pastoralists;
- To identify cultural factors that serve as climate change risk multipliers for Afar women;
- To identify gendered adaptation strategies to climate change among the Afar; and
- To suggest possible recommendations to effectively deal with climate change issues in the years ahead.

Review of Related Literature

Climate Change and Pastoralism

Rainfall variability between and within the years have made fluctuations in climatic conditions defining features of the drylands of Africa. However, pastoralism as a livelihood system enables people to cope with their harsh environment (Hesse and Kotula 2006). The most common characteristics of pastoral communities are that their livestock depend on natural resources while rainfall significantly determines the quality and quantity of water and grazing resources. Animals are predominantly indigenous species and provide extra-economic (social, cultural and religious) values, and the management of resources is based on common property tenures while access to grazing land and water depends on flexible and reciprocal arrangements based on negotiations (*Ibid.*).

Drought is frequently experienced in the pastoral areas in East Africa, which pastoralists regard as 'an act of God'. According to anecdotal evidence collected by Markakis (2004), drought cycles have become more frequent, shortening from ten to five years in the past and from five to three years at present. This directly affects the livelihood of pastoral and agrarian communities, as their income is directly dependent on rainfall (Diro, Thomas, and Len 2011). From the context of pastoral settings, drought implies two or more consecutive years when rainfall is less than 75 per cent of the long term average (UNDP 2000). Pastoralism all over Africa, as a system directly dependent on rangeland resources, is threatened by increased drought.

Climate change in Ethiopia, as in most sub-Saharan Africa, is manifested through extreme variability in rainfall and increased re-occurrence of drought[1]. The Afar pastoralists of Ethiopia have been repeatedly affected by recurrent droughts which occur every few years (Wolde-Georgis 1997; Helland 1980). For instance, Wolde-Georgis (1997) reported that out of the 19 infamous droughts and associated famines that took place in Ethiopia from 1543 - 1994, 17 of them involved the Afar pastoralists and pastoral areas. The loss is also very devastating. According to Helland (*Ibid.*), for example, in the great famine of 1973/74 alone, the Afar lost a fourth of their livestock and a third of the human population is believed to have perished.

Impacts of Climate Change on Pastoral Women

Although climate change affects both male and female pastoralists, the impacts are thought to be greater on women (Daniel 2005).

Based on Araujo and Quesada-Aguilar (2007), for example, women and children are 14 times more likely to die than men during climate change disasters. While women in the pastoralist societies of sub-Saharan Africa have historically been unprivileged, the marginalisation of pastoralists as communities in general has overshadowed the extremely degraded position of pastoralist women (Daniel 2005). Of the 1.3 billion people living below poverty line in the world, 70 per cent are women (Denton 2002). Studies indicate that 28 per cent of women in rural areas are moderately - to - severely under-nourished, and of the 49 per cent of people in sub-Saharan Africa living on less than 1 USD a day, over 60 per cent of them are women (UNDP 2009b). Owing to their greater dependence on natural resources for survival and their close association with pastoral rangeland resources, pastoralist women have specifically been vulnerable to the ills of climate change (recurrent drought) and the traditional inequalities with men are exacerbated by the consequences of climate change (UN-ISDR 2008).

In developing countries, women's susceptibility to the negative effects of climate change has been compounded by their low educational levels, as a result of which many of the warning information does not reach women or they do not understand them (Brody, Justina, and Emily 2008). In a study investigating the perception of women and men towards climate change, Dhanashri (2010) reports that, unlike men who have a humble knowledge of climate change, women were too unknowledgeable about it that it was difficult to elicit knowledge or perception of women on climate change.

Women's greater vulnerability to climate change can also be understood in terms of the numerous roles they play. Case studies conducted in South Africa and Nepal show that gender-differentiated impacts of climate variability were because of the unequal distribution of roles and responsibilities of men and women. Climate variability resulted into women bearing the most burdens resulting from climate variability impacts (Babugura 2010). Existing evidence suggests that the demands of the roles played by women pastoralists become even more challenging during drought (IPCC 2007; Wawire 2007; UNDP 2009a).With the increased re-occurrence of drought, pastoral women are burdened with more and more labour responsibilities.

To illustrate this point, during extreme droughts, when livestock losses have impoverished households, it is the woman who feeds the family by gathering wild grains (Katherine 2008). Drought also results in reduced amount of available water and women and children (mostly girls) will have to face the challenge of walking long distances to fetch water (IPCC 2007; UNDP 2009a). In addition, the responsibilities of women increase during drought because of increased mobility (increased transport labour and house construction), and increased labour for the collection of feed for animals (Lemlem and Yemane 2007).

Among the Borana pastoral group in Ethiopia,even less subtle changes resulting from recurrent drought, such as a shift from pastoralism to agro-pastoralism, has resulted in increased workload to women (Getachew 2007). Moreover, as migration due to climate change-induced disasters increases, so does conflict over land, water and other rangeland resources; and consequently increased sexual violence against women of the 'other' ethnic group (Reuveny 2007).

Gendered Adaptations to Climate Change

The most common pastoral adaptation practices for recurrent droughts are income diversification or involvement of members of the family in many extra-pastoral economic activities, migration, and so on (Nagar 2001; Brody, Justina, and Emily 2008). Pastoral supportive activities for adaptation to recurrent drought include agriculture, fishing, migration, use of wild fruits and dried plants as human and animal feed, herd splitting, selling of livestock, water harvesting and petty trading. These extra pastoral activities are based on the utilisation of household labour, thereby giving the traditional division of labour a new dimension (Nagar 2001). Women often do not have the opportunity to enter the formal workforce because of their time-consuming domestic responsibilities, lack of education, or certain social or cultural expectations and norms (Brody,Justina and Emily 2008) and are, therefore, likely to take a very limited range of jobs compared to men (Nagar 2001).

Nagar (*Ibid.*) argues that, as a result of drought, new forms of the pastoral way of life with new emerging norms and values have been formed. The new emerging values include reduced value of livestock as prestige and power indicators, reduced trust in the pastoral modes of life as sustainable livelihoods, and increased value attachment to cash income than the number of livestock and individuality (Nagar 2001; Songok *et al.* 2011). Economic pressure has forced pastoralists to abandon some hitherto very common practices such as incurring huge expenses for marriage in terms of livestock number (Dhanashri 2010), as well as to adopt strange values such as individualism (Fiona 2007), sedentarisation (Fratkin and Smith 2005) and labour migration (Meze-Hausken 2000; Reuveny 2007).

Women are not only passive victims of climate change but are also effective agents of change in relation to both mitigation[2] and adaptation[3]. Women's responsibilities in household and communities as stewards of natural resources has positioned them well to understand and innovate livelihood strategies adapted to changing environment (UNDP 2009b; Aoyagi, Suda and Shinada2011). Owing to their traditional responsibility of working with nature, pastoral women have many opportunities and challenges for innovative strategies of adaptation (UN-ISDR 2008). For instance, being responsible for most of the task of providing the household with its water supply, women and girls are the first to learn about declining water quality and quantity (Denton, Sokona and Thomas 2000).

Among the Afar pastoralists, women play an important role in adaptation to recurrent droughts. Theyhave been practising a lot of adaptation strategies ranging from collection of famine-time wild feed to petty trading, shift in the choice of animals, trees, cooking practices, and so on (Yemane 2004; Balehegn and Eniang 2008); and preserving dried meat (*Darru)* and processing ever-green wild food such as *Amberaarcu* and *Doberaglabra* and the like (Fassil, Diress, and Synnevåg 2001; Nagar 2001; Wawire 2000). However, current knowledge of adaptation and adaptive capacity is insufficient for making reliable prediction of adaptations, identifying specific areas where institutionalised support for adaptation can be rendered, rigorous evaluation of planned adaptation options, measures and policies (Barry & Olga 2000). Moreover, even if women have enough knowledge of their situation and have an emergency adaptation mechanism, they may not be aware of other options and strategies available for them. The provision of such options is only possible when what they know and understand is known (Barry and Olga 2000; Mitchell, Tanner and Lussier 2007; Mannke 2011).

Cognisant of the gendered nature of the issue of climate change, it is now widely accepted and recommended that successful adaptation to climate change will require an acknowledgement of the gender dimension of climate change *(Wamukonya and Skutsch 2002; Valerie et al.* 2002; Gurung and Mwanundu 2006; Bäthge 2010). A gendered approach to climate change requires more than a set of disaggregated data showing that climate change has differential impact on women and men. It requires an understanding of existing inequalities between men and women and of the ways in which climate change can exacerbate these inequalities.

Conversely it also requires understanding of the ways in which these inequalities can intensify the impacts of climate change for all individuals and society (Brody, Justina and Emily 2008); and taking into consideration such inequalities in developing climate change adaptation and mitigation policies (UNDP 2008); identifying the gendered inequality in vulnerability to climate change (Dankelman 2002; Denton 2002); and gender mainstreaming[4] of any strategy or plan of action implemented to improve adaptation to climate change. The participation of all members of society, including men, women and children, is vital to sound decision-making and to successful adaptation that promotes equitable human development and the achievement of Millennium Development Goals (Katherine 2008). Women's knowledge and practices should be documented and shared to demonstrate how addressing issues of gender equality and empowerment can strengthen climate change adaptation (Carvajal-Escobar, Garcia-Vargas and Quintero-Angel 2008).

Methodology

Description of the Study Area

The Afar people are Cushitic-speaking people living in the arid and semi-arid areas of Ethiopia, Eritrea and Djibouti. The Afar pastoralists are the second largest pastoral communities in Ethiopia (Yemane 2004). According to the Central Statistics Authority (CSA 2007),

AfarRegionalState has a total population of 1,411,092 persons, of whom 624,754 are women. Less than 14 per cent of the population lives in urban areas (*Ibid.*).

The Afar Region of Ethiopia is structured into five zones and 29 districts (known as *weredas*). Historically, the economy of the Afar is based on multi-species livestock husbandry. But with increased vulnerability to drought and famine due to ecological disasters, they now depend mainly on camel and goat pastoralism (Wolde-Georgis 1997).

The study was conducted in *Aba'ala* district of Northern Afar Administration which was selected based on its severe proneness to recurrent drought and famine. The northern Afars are mainly agro-pastoralists who combine animal husbandry with marginal agriculture. Their settlements are semi-permanent – situated near permanent water sources and small trading centres. Since rain-fed agriculture alone is often not promising, they employ water diversion techniques as well to grow their crops.

Methods of Data Collection and Analysis

The study relied on a wide variety of primary and secondary sources. Primary data were collected from ordinary pastoralists of different socio-economic backgrounds (stratified based on age and gender) as well as from administrative bodies at various levels. Secondary data were obtained from previous studies, official reports and other relevant documents.

The study employed mainly qualitative research methods although some level of descriptive quantification was undertaken based on a sample of respondents using a structured interview schedule. Qualitative data were obtained using ethnographic methods. The specific tools employed were Key Informant Interviews, Focus Group Discussions and Case Studies. The qualitative data were analysed and presented in narrative forms. Results of the case studies are integrated into the major findings as supportive illustrations.

Structured Interview Schedule

Individual structured interviews were held with purposively sampled 40 women and 40 men selected on the basis of variations in age and socio-economic brackets. Such individual interviews involved the study of differences in perception and attitudes towards recurrent drought and investigation of the gendered adaptation mechanisms and strategies employed.In the individual interviews, the raw narrations on women's inherent inequality in Afar and its consequences in drought adaptation were inductively categorised (Auer-Srnka and Koeszegi 2007) and quantified.

Key Informant Interviews

Clan elders, opinion leaders, and informed key communicators were approached for a series of interviews aimed at obtaining village-level data on the modes of livelihoods, major events in the area across time and overall trends in climate change, as well as local practices and perceptions related to drought and coping. A total of 30 key informants (15 clan leaders and community elders, 10 elderly women, and 5 government workers) were selected on the basis of age, eloquence, as well as experience and wisdom.

Focus Group Discussions

Focused group discussions were conducted with clan leaders, elderly women, and administrative bodies. There were a total of 40 focus group discussion participants. Each group was composed of 6-8 persons. At the outset, homogeneous groups based on age and sexe were organised. At later stages, people of different socio-economic brackets were treated together. A typological categorisation of different narrations (Auer-Srnka and Koeszegi 2007) about each and every traditional institution (stories, events, by-laws and so on), were made and each category was coded in order to quantify the gender sensitiveness of each traditional institution. With these focus groups, using the Harvard Analytical Framework tool, the different pastoral and extra pastoral roles of women, men, boys and girls were differentiated and informants were asked to rank the roles of the different categories using the traditional *gebeta* game, a variant of the East African *bao* game used for other similar ranking exercises (Franzel *et al.* 2001).

Case Studies

In order to substantiate major findings and arguments of the study, 10 case studies (personal histories) were collected using an in-depth interviewing technique. This was conducted with women household heads that have distinctive success and failure stories to tell regarding climate change impacts and adaptation. The specific women were selected based on recommendations from local residents. Only a few cases are presented in this chapter as illustrations.

Results and Discussions

Demographic and Socio-economic Characteristics of Respondents

In terms of sex composition, half (50 per cent) of those considered for quantitative interview schedule were women. For comparative purposes, young and old people from both sexes were considered for the study. Hence, half (50 per cent) in each sex category fell within the age range of 18-25 while the remaining half (50 per cent) were above 40 years of age. In terms of education level, the majority (85 per cent) of them were not able to read and write. The remaining (15 per cent) had at least basic education. Most of them (75 per cent) identified themselves as poor, while 20 per cent and 5 per cent were categorised as medium and rich, respectively.

Climate Change Knowledge, Perceptions and Attitudes

Until half a century ago, Afar elders traditionally divided the year into seven different seasons, called: *Segum*, *Karma*, *Konayto, Datrob, Debaba*, *Zeza'e* and *Gilal*or *Hagaya*. Today, such a meticulous classification is non-existent and hence, some of these seasons are not known particularly to the younger generations. As indicated in Table 1, most young men and women did not mention *Konayto, Datrob*, *Debaba* and *Zeza'e*, while almost everyone mentioned *Segum*, *Karma* and *Gilal*. Results of the interviews indicate that men tend to be more knowledgeable about the different seasons, as they mentioned more seasons than women. Similarly, older people were found to mention more seasons than younger ones. The elders claim that the inability of the younger generation to identify traditionally obvious seasons is mainly due to the increased detachment of the young generation from herding and herding-related activities, which is usually considered an evidence of the continual regression and degeneration of Afar normative culture which was perceived to be one of the factors for deteriorating living conditions.

Table 1: **Knowledge of seasons among Afar pastoralist, by sex and age group**

SN	Traditional Seasons	Percentage of Respondents who mentioned the seasons			
		Young Men (18-25) (n=40)	Old Men (≥40) (n=40)	Young Women (18- 25) (n=40)	Old Women (≥40)(n=40)
1	Segum	78.4	100	86.6	94.5
2	Karma	84.6	89.9	94.7	82.7
3	Konayto	33.1	42.3	9.3	12.2
4	Datrob	2.5	15	3.3	4.9
5	Debaba	13.4	28.9	2.8	9.6
6	Zeza'e	22.1	53.7	20.4	33.6
7	Gilal(Hagaya)	89.1	100	98.2	99

Source: Authors' own survey

Afar informants asserted that the amounts of rainfall has decreased steadily from time to time especially in the last 4-5 decades. Drought, which is a common phenomenon for the Afar, is identified by three different Afar words with the same meanings: *Abara*, *Adalsa*, and *Hagaya*.

The Afar have been so exposed to various drought occurrences that they have developed nomenclature and categorisation of different types of drought that occur at different times. They have also developed indigenous ways of predicting drought occurrences. It is important to note that Afar pastoralists used to be very precise with regard to their knowledge in the timing of the different seasons. This is mainly because their livelihood, which was based on moving animals from place to place, depended on the strict knowledge of timing of dry and wet seasons. Today, not only has the amount of rainfall decreased, but the timing and usual spatial distribution has also been greatly disturbed. This is demonstrated in one of their prominent proverbs, "Rain discriminates between the two horns of an ox." One elder described the current rainfall situation in his area as "rain which does not fall on either of the two horns of an ox."

According to the qualitative interviews, there is a consistent agreement among all informants as to the reason for increased drought. Informants believe that the increased drought and its negative consequences are a "curse from Allah"- a price people pay for all their behavioural transgression and disobedience to heavenly commandments. Elders explained that in the past, people used to be so united that almost all Afar individuals were able to survive the hardest of times, as resources were being shared among clan and family members. However, now individualism is replacing the traditional values of communality in Afar, and this is believed to have caused God's wrath. Moreover, in the past, it was not uncommon for families and individuals to prepare a feast of meat and milk, and invite community elders and sheiks to their houses as a sign of respect and beg them for blessings (*Du'a*). Similarly, the Afar used to practise extensively the religious traditions of *Zaka*, where a rich herder shares milk, meat and sometimes even live animals with poorer families. Now such practices are rare partly because of resource scarcity and impoverishments. The Afar believe that it is such 'unfavourable' social changes that are angering Allah and making him hold back the rains.

Exceptionally, a small proportion of informants associated drought with over-population (both in people and livestock), over-grazing, and deforestation of rangelands which, in turn, cause scarce and erratic rainfall patterns. Scientific knowledge on climate and climate change impacts among women informants were found to be limited as compared to that of men, mainly due to the lack of exposure of women to public media and information sources. However, women tend to be aware of the steady reduction in the amount of rainfall and the re-occurrence of drought. People, however, do not directly talk about drought in its meteorological sense; rather they use a lot of bio-physical indicators in order to describe the re-occurrence of drought. These bio-physical and social phenomena are related usually to livestock productivity, bio-physical changes, and socio-economic and cultural changes (see Table 2).

Table 2: **Proxies of climate change in Afar**

SN	Indicators Related to Livestock	Indicators Related to Bio-physical Environment	Socio-economic and Cultural Changes
1	Reduced milk yield in camel and cattle	Drying up of ancestral water wells*	Women in towns trading
2	More lean animals	Failure of regeneration of some tree species, and increased proportion of others	More people migrating to Saudi Arabia
3	Increased number of goats and camel and less number of cattle	Increased flood and soil erosion	More conflicts with neighbouring highlanders
4	Less number of animals with '*ala*' or rumen fill	A shift in the type of crops sown, as some are not viable any more	Increased individualism
5	Repeated livestock pandemics	Drying up and reduction in the amount of flow in perennial rivers	Women no more selling ropes made from sisal plants (*Agave sisalana*), as most of the sisals are unable to grow under current drought conditions
6	More cattle and camel on the mountains than on the flat rangelands	Reoccurring volcanic eruption**	
7		Reduced number of sightings of wildlife species:example; antelopes, gazelles, lions and leopards,	

* In the course of time, due to recurrent droughts, many ancestral water wells identified with different families and clans in Afar have disappeared. Participants of the focused group discussions identified the following water wells around Aba'ala town that dried completely for the first time in history: *Baiela* (located around *Erepti* – dried two years ago), *Dardarela* (located around *Sasu* – dried four years ago), (*Asgahartuli,* located around *Wegare*-dried 6 years ago).

**Recent volcanic activities around *Ertaela, Dabah, Duraele, Gralaele,* which have been reported extensively in the mass media and scientific forums, have been associated with drought by the Afar.

Source: Authors' own survey

Climate Change Indicators and Impacts

The Afar describe drought indicators in terms of the changes to the bio-physical environment and livestock conditions. As indicated in Table 2, changes to the environment are related to the loss of bio-diversity and declining supply and sustainability of pastoral resources. Drought has also severely affected the physical condition of livestock and their productivity.

Impact on Livestock

Pastoral livelihood is dependent on livestock; as a result, pastoralists are very keen observers of direct and indirect changes to their livestock. Elders explained with a deep sense of nostalgia how a single camel used to support a family of ten, and how their cattle and camels used to graze only some metres away from home retuning home with rumen fill (*Ala*) after just a few hours of grazing and browsing.

Drought-induced changes to livestock include reduced body condition, fertility and increased weaning time of different animals. The number of animals per household is also said to be greatly reduced. While camel and cattle ownership decreased, sheep and goat have become more abundant. This is mainly because it is relatively easier to rear small-stock animals under extreme fodder scarcity conditions. Moreover, sheep and goat, unlike camel and cattle, are able to utilise mountainous rangelands, which were not traditionally common grazing places for the Afar. Table 3 provides information on the extent of drought-induced changes in livestock and livestock-related characteristics for the last fifty years.

Table 3: **Effect of recurrent drought on livestock productivity**

Livestock-Related Indicators of Climate Change (Recurrent Drought)	**Comparison**	
	50 years ago	Current
Number of People supported by one lactating cow, throughout lactation time	6-10	3
Number of times a lactating camel was milked per day	3	2
Average number of camels per household	20-50	3-10
Average number of cattle per household	50-300	3-10
Average number of goats and sheep per household	50-300	100-300
Average number of inheritance children get from their parents at birth	5-20	0-3

Source: Authors' own survey

Impact on Bio-physical Environment

Over the past few decades, there has been a significant change to the landscape of the Afar as a result of recurrent drought and other anthropogenic factors. Afar informants indicated that the species composition and the overall vegetation cover in Afar rangelands have substantially decreased. Recurrent drought is repeatedly indicated to be the main reason for the observed decrease in vegetation cover, while cutting trees for construction and fuel wood has also been identified as an important factor. The uniform occurrence of drought throughout Afarland has reduced the need for mobility as a result of which people have decided to settle in small villages. Sedentarisation in turn resulted in over-grazing and huge pressures on forest resources for fuel.

Those pastoralists who have been actively engaged in crop cultivation besides animal husbandry also indicated that they were not able to grow the crop varieties they used to grow before. For example, two specific varieties of sorghum and maize (*Gedalit* and *Humera)* and other varieties for *Teff* and barely have been completely abandoned because the amount of rainfall is not enough for these varieties to grow.

Impact on Social Relations

With increasing trends in climate change and drought, the historically rooted resource-based conflicts and animosity with neighbouring highland farmers have increased in frequency. Ethnic conflicts have adverse impacts on pastoralists not only directly through damages caused to human lives, but also through the temporary restriction to market interaction with neighbouring cultivators. Moreover inter-ethnic and inter-clan conflicts usually result in the abandonment of grazing land ultimately leading to resource wastage. However, respondents indicated that although incidents of conflict have increased, their impact on the lives of people and cattle has decreased due to the heavy involvement of the police and national defence forces.

Gendered Impacts of Climate Change

The Afar exercise sedentarisation as a form of economic adjustment in response to climate change and drought. Sedentarisation could facilitate women's access to modern facilities like piped water, education and health (Fratkin and Roth 2005). However, sedentarisation is said to increase the labour expenditure of women (Fratkin and Smith 2005), mainly due to the increased role of women in preparing, transporting and selling of marketable household products, such as milk, curd, butter and fuel wood, which make up the main livelihood of a settled pastoral household. According to informants, sedentary life brings more responsibilities and workload to women as most children are likely to attend formal schooling, leaving adult women behind to take care of all household chores. However, women in Eka, and Dinemellivillages, which are the nearest to Aba'ala town, mentioned that the provision of piped water within their premises has minimised the risk of fetching water from faraway places which also saves their time and energy. Women in the peripheral villages of Asangola, Haridan, and Kala, still have to walk for 3-7 hours to fetch water. With the current drying up of ancestral wells and other water bodies due to excessive temperature and drought, fetching water for household consumption has become a big challenge for many Afar women in rural areas.

Women in Afar traditionally prepare ropes and sacks from the local sisal plant (*Agave sisalana*) and sell it to highland salt traders in order to buy some food grains for their families. However, with the advent of severe droughts, there is a loss of bio-diversity and hence the plant which provided the raw material for rope-making is fast disappearing. Besides, perennial rivers which regularly provided water for the preparation process are drying up early. This results in loss of income. Similalry, Buchler (2009) has also reported loss of traditional income sources for women due to problems of vegetables and crop production as a result of climate change in SnoraMexico. The following (Case 1) illustrates how Afar women suffer most during drought periods.

Case 1

Halima Mohammednur, a young woman aged 30, was born in Haridan Village. She got married when she was 15. Her marriage lasted for five years as her husband suddenly decided to divorce her apparently because she was not able to produce children. Although divorcing a woman just because she is barren is against the *Adda* and the teachings of Quran, the couple was finally separated. After divorce, Halima returned to her father's house and served the household with a lot of responsibilities. She spent much of her time fetching water, supporting livestock and preparing ropes to be sold at Aba'ala market. During drought years, Halima had to walk for more than ten hours a day to fetch water for the household and young animals; moreover, she had to carry grains for milling in Ab'ala town which is located at a distance of 26km from her own village. She desperately looked forward to a time when such facilities would be available in her small village.

On the positive side, however, one notable outcome of recurrent drought in Afar is a breakdown of traditional norms related to gender relations. Pastoral women challenged the usual ideology of women's economic dependence on men as the sole bread-winners. More and more women now dare to engage in other non-pastoral income-earning activities. Impoverished women household heads have initiated micro and small enterprises (besides cultivation) as viable livelihood options to augment their ever-declining pastoral income (Boruru *et al.* 2011). Although resorting to this mode of life has not been traditionally preferred by the Afar, it has obviously improved the living conditions of those women who are practising it.

Aspects of Afar Culture as Climate Change Risk Multipliers

According to Afar tradition (*Adda*), women are subject to unfavourable treatments in society with numerous restrictions and prohibitions. Their participation in community affairs is limited. Women are considered "unclean" and their interactions with people (and animals) are sometimes restricted. Their access to good quality and quantity food is also generally at stake as priority is always given to men and boys. This is further complicated during drought where resource scarcity squeezes down women's share even further.

Another cultural factor which has a risk multiplier effect for Afar women is the tradition of property and asset inheritance (*Warsa)*. Women in Afar are not entitled to any kind of wealth, apparently even to the one that they earned and produced. The inequality in wealth acquisition starts at birth. Although there is a long standing tradition of assigning animals to children at birth, female children are either totally excluded or they receive only half that of their male siblings. Such differential treatment of men and women in inheritance rights continue into adult life too. In case of divorce, women are not traditionally entitled to a share of the household assets although such practices are now gradually being diluted by *Sharia* laws which allow women to take only a third of the household asset during separation. This inequality in the inheritance pattern and wealth distribution has a profound effect on the ability of women to survive drought and related stress.

Similarly, the Afar have a tradition of widow inheritance (*Hixu),* following the death of a husband with the aim of supporting the wife and children of the deceased, and wealth sharing by kinsmen of the dead person. Often the woman and her young children will be left to survive almost on nothing as the new husband and older male children scramble for the animals in the household, often making them physically and morally weak to cope with drought.

Another marriage pattern widely exercised among the Afar is *Absuma* which gives uncles a full right to claim marriage with their nieces. If a girl refuses to get married to her uncle, the man requesting marriage will put a prohibition (*Aqdi*) preventing the girl from marrying someone else. Even more problematic is that even if the uncle who put the *Aqdi* dies, the girl will not be free of the prohibitions, unless the dying man speaks of dropping the *Aqdi* to a respected elder before he dies. Therefore, even after the death of the uncle, no one will dare to ask the girl for marriage for fear of offending the clan and family members of the deceased uncle. The problem with the tradition of *Absuma* and *Aqdi*is not only that the girl is forced to marry against her will, but that most of the time the uncles are older than the girls by about 20-50 years. This big age difference means that the man already has another household to care for, and that he is marrying the new wife only to produce more children. Most womenmarried to their old *Absuma*, are by far highly exposed to the negative effects of drought than those married to a young man in the usual way.

Under-age marriages are also common in Afar. The average age at marriage of the 80 interviewed women was 14.36, which is obviously a very young age for a girl to get married. Under-age marriage by itself may not directly affect the vulnerability of women to drought or their capacity to adapt; however, it does make women to give birth to large number of children at an early age which in turn restricts their capacity to stay active in extra-domestic activities to support their households. Of the 80 interviewed women, 68 had 3 children within five years after marriage. Under-age marriage is obviously a problem for the girl being married; however, it is difficult to convince parents because they justify under-age or early marriage as a way for avoiding pre-marital sex and pregnancy outside wedlock (Boyden,Pankhurst andTafere 2012).

Another area of disparity between men and women among the Afar is pertinent to the culturally assigned gender-based classification of roles and responsibilities. Despite the subordination of women by men, and the lower status they hold in Afar society, a close examination of the household division of labour clearly reveals that women actually play larger roles than men in running and maintaining the household. However, this is not well recognised. During drought times, household labour requirement is higher than normal years but women shoulder much of the responsibilities of supporting the family (buying and selling, milling, transporting food and other items, and so on). Even during famines, when people are hard hit by food shortages, it is the responsibility of the woman to collect unconventional famine food and save the family from starvation and death. This is because it is considered to be too low a task for a man to be involved in the gathering of any kind of food (either grains or wild food items). Table 4 shows a typical gender-based division of labour.

Table 4: **Household labour classification among the different genders**

Activity (Chores)	**Responsibility Ranking**			
	Boys	Men	Girls	Women
Preparation of food for household consumption	0	0	1	4
Fetching water for domestic use (people and young animals)	0	0	2	3
Irrigating crops in flood cultivation, and irrigation	1	3	0	1
Preparation of tea, coffee and other leisure time foods and drinks	0	0	1	4
Purchasing and transporting grains from market to miller and then home	0	0	0	4
House construction and transporting the house during movement	0	0	0	5
Looking after children (all tasks of babysitting)	0	0	2	3
Fetching fodder for young and lactating animals	1	3	0	1
Preparation of ropes from Agave plants for selling to salt traders	0	0	0	5
Preparation of milking utensils (*karfo*)	0	0	0	5
Milking sheep and goat	0	2	0	3
Milking of camel	0	5	0	0
Preparation of curd, and butter	0	0	0	5
Herding goats	3	0	1	1
Herding camel	1	4	0	0
Transhumant migration with dry animals	2	3	0	0
Participating in community activities, and decision making (*Adda*, conflict resolution, *Edo (range scouting)* etc.)	0	5	0	0
Cleaning and maintain house, animal barn etc.	0	0	2	3
Preparation of *Oloyta*(traditional bedding made of plant leaves)	0	0	0	5
Preparation of dried meat as drought time food	0	0	0	5
Collection and processing of famine time wild foods	0	0	3	4
Total Responsibility Scores	8	25	12	61
Percentage Household labour share	7.6	23.8	11.4	58.1

Source: Authors' own survey

According to *Gebeta* ranking of exhaustive list of all household chores listed in Table 4, women contribute 58.1 per cent of household labour share, while men contribute 23.8 per cent, nearly half that of women. Similarly girls contribute more than boys do. With increased drought incidents, women tend to do more laborious jobs than men.

Climate Change Adaptation

The Afar pastoral communities, through long-time interaction and experience with recurrent drought, have demonstrated a tremendous ability to adapt to climate variability. They often employ sophisticated integrated (modern and indigenous) and continually evolving processes and practices to respond to risks which also help them take advantage of new opportunities.

Indigenous Weather Forecasting

Owing to the importance of predicting weather conditions for livestock mobility and other pastoral practices, the Afar pastoralists have developed a number of traditional weather forecasting techniques. Women play important roles in some of these traditional practices. There are prominent magicians in the neighbourhoods who are involved in forecasting of climatic events in addition to their routine role in solving subtle problems in the day-to-day life of ordinary village people. Often their weather-related forecasts are based on observation of bio-physical entities, such as insects, plants, wild animals, clouds, stars, wind direction, smell of the soil, and animal footprints. Despite the age-old adaptive values of such practices, they are nowadays being discouraged by religious leaders (*Sheiks*), who argue that forecasting of future events lies within the discretion of God (Allah) only. Table 5 indicates the common practices and traditional techniques of weather forecasting as explained by three key informants who used to be traditional magicians.

The importance of the indigenous weather forecasting activities to the Afar pastoralists cannot be underestimated as it serves as the basis for household decision-making on important pastoral activities like mobility, de-stocking, marriage and so on. Such traditional practices help pastoralist people make preparations for upcoming bad seasons. The common preparations and activities undertaken as proactive measures against drought are:

- Dispatch of traditional range scouts (*Edo*) for assessment of rangeland resources in different localities before the next move can be made;
- Collecting information (*Dagu)*[5], about social, economic and political developments in the villages in response to the imminent danger (drought);
- Asking (begging) well-to-do relatives and clan members to temporarily adopt children, and move animals to safer areas (especially those that are likely to be affected by the looming drought);
- Selling of male animals (oxen and male goats), and buying male camels for transportation purposes during mobility;
- Converting cattle into money and purchasing grains from markets;
- Sending adolescent sons and male animals to distant rangeland sites;
- Slaughtering male animals (especially those that are emaciated) and storing dried meat for future use; and
- Collection of famine foods, which includes the gathering of wild fruits from such trees as *Zizyphusspina-christi* (*Kusra*), *Cordiaovalis* (*Medera*), *Balanitesrotundifolia* (*alayto*), *Doberaglabra* (*garsa*), *(Humerto)*, *Grewiavilosa* (*Hibeli)*, *Grewiaerythrea(Hida)*, and*Yalue*(species not identified).

Table 5: **Indigenous weather forecasting in Afar**

SN	Category	Observation	Weather Prediction
1	**Traditional Astronomy**	'*Dohra*', a star appearing in the east around 8:00 in the evening and fades at 4:00	Some rain
		'*Dohra*' appearing in between west and east (middle of the sky) at night	No rain (drought)
		'Dohra' goes from east to west	Very good rain
		A group of three stars called '*Hotali*' appearing at around five in the morning	Very good rain
		A star named '*koimat*' spotted by a man milking a camel	Very good rain
		Red coloured sky to the east early in the morning	No rain (drought)
		Red coloured sky in to the north early in the morning	Good rain
2	**Bio-physical observations**	*Doberaglabra* tree too green during dry season	No rain (drought)
		Doberaglabra tree shedding leaves in the dry season	Good rainy season
		Camel, wild animals (antelopes and gazelles) mating	Good rainy season

		Reduced mating of camel and wild animals during the normal breeding season	Drought
		Large birds flying to the north	Good rainy season
		Large birds flying in direction other than north	Drought
3	**Social Changes**	New people and cattle coming to nearby rangelands	Drought
		Increased conflicts over animals, and petty things	Drought
		Increased price of grains and reduced price of animal	Drought

Source: Authors' own field data

Gender-Based Adaptation Strategies

The adaptive capacity of the Afar differs considerably among men and women, and depending on their level of exposure to modern facilities such as education, tap water, transportation, and so on.

In most of the reported cases, women from Aba'ala town were more likely to have employed new strategies for adaptation than those in villages. The common adaptation mechanisms include strategic labour migration decisions, shrewd management and resolution of conflicts related with competition over pasture and water resources, rapidly changing agrarian and pastoral practices, as well as complex socio-economic linkages between rural and urban areas.

Women play a vital role in enabling the family to survive drought and drought-related shocks. Some recent economic roles played by women tend to contradict with the provisions of the customary norms of the Afar. For example, Afar women have now started to move out of their villages to embark on independent economic transactions with neighbouring highland communities. Although this practice is not well appreciated by men, it gives women an opportunity for income diversification and is immensely valued particularly by women household heads.

Additional activities that facilitate drought adaptation and coping include food aid and other assistance packages by governmental and non-governmental organisations. Men and women benefit from such packages although men tend to believe that food aid is meant for women. It is often regarded a shame for men to wander around for food whatsoever might be the case – be it receiving from aid distribution centres or even getting it from the markets.

However, a lot of factors set a limit to the capacity of women (and men) to adapt to the recurrent drought as can be explained below.

Firstly, Afar pastoralists strongly believe that recurrent drought is the result of a curse from God, and hence virtually nothing can be done to reverse the situation.

Secondly, most women (and men) lack the necessary information and experience as to what can be the best practice of adaptation.

Thirdly, Afar cultural norms and value systems put a limit to the extent to which women can innovatively explore adaptive strategies to improve their living conditions. As explained in the previous section, successful adaptation strategies such as mobility and petty trade by women have continuously been discouraged by Afar culture (*Adda*), as a result of which many women refrain from taking up these activities for fear of rejection and social disapproval. Table 6 is a presentation of the gender-differentiated drought adaptation mechanisms and respondents' ranking of the contribution of men and women.

Table 6: **Gender-differentiated drought adaptation strategies**

Adaptation Strategy		**Practised by**	
		Men	Women
Extra-Pastoral Activities	Petty trading		*******
	Labour migration to Saudi Arabia, and other Arab countries	****	
	Working in government offices	*****	*
	Education, to earn a living outside the pastoral sector	**	*
	Fattening of small stock (sheep and goat) for market		***
	Collection, processing and cooking of aid food		*****
Traditional (pastoral) Coping Strategies	Preparation of dried meat	**	*****
	Collection of famine wild food		*****
	Mobility to areas of relative fodder abundance	*****	**
	Selling animals to purchase grains	***	***
	Herd splitting	**	
Pastoral Structural Adjustment	Sedentarisation	***	***
	De-stocking	***	***

Note: Number of asterisks indicates the level of responsibility of each sex, as scored during FGDs.

Source: Authors' own field data

As can be observed from Table 6, there are pastoral and non-pastoral adaptive mechanisms developed by Afar women over a period of time which need to be properly harnessed and refined in order to keep them resilient in a way they respond to shocks. The following case (Case 2) demonstrates how some Afar women pick up petty trade as an alternative livelihood strategy in response to drought.

Case 2

Hasna Ahmed Abrahim is a 50-year-old woman born in Kala. She was married when she was 16 to a man from a very rich family. She gave birth to five children, and then got divorced from her husband. She decided not to marry again and started to look for ways to raise her young children. Despite the restrictions against women of publicly trading in the markets, she started to buy grains in the highland markets and then sell it in the Afar area (Aba'ala). After years of trading, she managed to buy two camels, which she sent to her relatives in the village; the camels gave birth to many other camels, and now she owns about 15 camels. Her trading activity in the town also went so well that she managed to build an eight-room villa in Aba'ala town, a big achievement for an Afar woman. By becoming financially strong, she also helped her daughters get married to people from rich families.

Conclusion and Recommendations

Conclusion

The last 4-5 decades were crucial times for Afar culture, livelihood and their pastoral mode of production. Drought, coupled with encroachments and resource-based conflicts with neighbouring ethnic groups have resulted in huge livestock loss and the subsequent impoverishment of the people.

Current trends in climate change, coupled with other social, political and demographic factors, have made pastoralists even more prone to drought and famine over the past few decades. Drought has never been new to Afar pastoralists but its occurrences have changed from one of cyclic to irregularly frequent nature. An increase in the magnitude of climate change impacts in recent years has significantly incapacitated pastoral innovation on adaptation and coping.

Afar women are worst affected during drought compared to men because of a multitude of factors:

- Women generally have limited access to climate change-related information and whatever information disseminated is not so well comprehensible to them given their low level of literacy.
- Women also have diminished access to food while, paradoxically, sparing additional labour to fulfil the responsibilities of family care and animal tending.
- The cultural subordination dictated by the patriarchal social organisation, further exacerbates differential impacts of climate change on women.

Despite such impacts, the Afar have long developed successfully tested adaptive strategies against environmental shocks and vulnerability through effective management of their resources. When old coping strategies failed to provide adequate solutions to environmental stress, they flexibly designed new forms of adaptation in response to changing circumstances. Older strategies include the establishment of strong economic and social support networks; establishing institutions for grazing land management, herd splitting, herd diversification; and resorting to small-stock animals to promote quick convertibility to cash in times of crisis. Relatively newer adaptive mechanisms include, among other things, resorting to agriculture and sedentary life, trade and wage labour migrations.

Despite their disadvantaged position with regard to climate change impacts, Afar women are better situated in the process of adapting to drought conditions. Women quickly learn and observe changes to the natural resource base. They can easily identify and make use of famine foods and undergo their own cooking practices. Besides, they are adepts in selecting appropriate drought-resistant animal species and have now started to innovatively use folk tradition to break the prejudice against women's involvement in petty trade and related activities. However, their low access to education and persistent cultural influences present the biggest challenges to effective climate change adaptation by women.

Recommendations

Based on the findings of the study, the following recommendations can be made:

- New livelihoods options, such as trade, labour migration, and education can be used as a means of securing a safe future for pastoralists. Hence, proactive measures should be taken to encourage and strengthen the capacity of pastoralists, especially women, to effectively adapt to climate change impacts.
- The lack of strong national policies on climate change points towards the need for enhanced political commitment by the state to integrate climate issues in development planning. The empowerment of pastoral women should be central to policies, programmes and strategies targeting climate change impact adaptation and mitigation.
- Successful adaptation to climate change in Afar cannot be realised without dealing with salient problems of pastoral women. Combining climate change adaptation with economic empowerment of women creates a window of opportunity for building mutual synergy on the path to poverty reduction and development. Hence, all stakeholders (community, governmental and non-governmental organisations as well as research institutions) should join hands in achieving this goal in a participatory and transparent manner.
- Climate change impacts cannot be properly addressed in the absence of strong functional integration between formal and informal institutions. Hence, the government should approach clan leaders to search for ways in which traditional *Adda* can be integrated with formal governance structures with the aim of identifying effective interventions against gender inequality, thereby setting a limit to its risk multiplying effect.
- In the long run, there must also be ways in which climate change and its gendered impacts can be incorporated into the curriculum of formal education in schools.

Notes

1. Drought, in this paper is defined as the periodic absence or decline in rainfall. Drought is a regular feature of the African climate, but it occurred with greater frequency in the second half of the 20th century compared to the first, reported in East Africa in 1960–61, 1968–69, 1974–76, 1979–81, 1991–93, 1996, and 2000. Severe famine occurred in 1968–73 in the Sahelian countries of West Africa (Burkina Faso, Chad, Mali, Mauritania, Niger, Nigeria, Senegal, and Sudan), and in 1982–84 in the Horn of Africa (Djibouti, Ethiopia, Kenya, Somalia, and Sudan).
2. Due to the scientific nature of climate change mitigation (reducing greenhouse gases), there are very little opportunities of mitigation for pastoral societies. However, pastoral societies can play a role in environmental rehabilitation by either reforestation programmes or adoption of proper utilisation of rangelands (Alyson, *et al.* 2008). However, this study is not concerned with pastoral climate mitigation strategies.
3. Adaptation refers to adjustments in ecological, social, or economic systems in response to actual or expected climatic stimuli and their effects or impacts. It refers to changes in processes, practices and structures to moderate potential damages or to benefit from opportunities associated with climate change (Barry & Olga 2000).
4. Gender mainstreaming is the process of assessing and incorporating the impact of any planned action on both men and women, including legislation, policies or programmes and ensuring that the concerns and experiences of both men and women are an integral dimension of the design, implementation, monitoring and evaluation of policies and programmes (UNDP 2009).
5. *Dagu* is a traditional information communication network in Afar. Whenever two Afar men meet anywhere, they first exchange the customary greetings, and then sit down in a nearby shade to converse about the weather and political condition of their respective places. Information communicated at *Dagu* is very credible because the Afar strongly believe that it is against the *Adda* of Afar to communicate wrong information. Therefore, whenever someone is communicating secondary information, then he is expected to indicate his sources.

References

Aoyagi, M., Suda, E. and Shinada, T. 2011. Gender inclusion in climate change adaptation. ADBI Working Paper 309. Available from:http://ssrn.com/abstract=1926494 or http://dx.doi.org. (accessed on 20th September, 2011).

Araujo, A. and Quesada-Aguilar, A. 2007. Gender equality and adaptation. Available from http://www.genderandenvironment.org/admin/admin_biblioteca/documentos/Factsheet%20Adaptation.pdf. (accessed on 14th August, 2011).

Auer-Srnka, K. J. and Koeszegi, S. 2007. From words to numbers: How to transform qualitative data into meaningful quantitative results. *Schmalenbach Business Review*, Vol. 59, January 2007.

Babugura, A. 2010. Gender and climate change: South Africa case study: Heinrich Böll Foundation – Regional Office Southern Africa.Available from:http://www.za.boell.org/downloads/gcc_south_africa.pdf. (accessed on 12th August, 2011).

Balehegn, M. & Ediang, E. 2008. Assessing indigenous knowledge for evaluation, propagation and conservation of indigenous multipurpose fodder trees: towards enhancing climate change adaptation in Northern Ethiopia. In *Traditional forest- related knowldege and sustainable forest management in Africa: Papers from the conference held in Accra, Ghana, from 15-17 October 2008.* edited by John, A. Parrota,Alfered Oteng-Yeboah, Joseph Cobbinah , IUFRO World Series volume 23.IUFRO (International Union for Forest Research Organizations)Vienna, Austria, 2009, pp. 39-46.

Barry, S. & Olga, P. 2000. Adaptation to climate change in the context of sustainable development and equity. Available from: http://www.ipcc.ch/ipccreprts/tar/wg2/pdf/wg2TARchap18.pdf. (accessed on 14th August, 2011).

Bäthge, S. 2010. Climate change and gender: Economic empowerment of women through climate mitigation and adaptation? The Governace Cluster. Available from: www.gtz.de/gender. (accessed on 4th November, 2011).

Boruru, E.O., Edward Ontita, William Okelo Ogara, and Nicholas Otienoh Oguge. 2011. Climate change and the emergence of helter-skelter livelihoods among the pastoralists of Samburu East District, Kenya.In *Experiences of climate change adaptation in Africa,* edited by Walter Leal Filho, Springer-Verlag Berlin Heidelberg, pp.97-110.

Boyden, J., Pankhurst, A. & Tafere, Y. 2012. Child protection and harmful traditional practices: Female early marriage and genital modification in Ethiopia. *Development in practice,* 22, 510.

Brody, A., Justina, D., and Emily, E. 2008. Gender and climate change: Mapping the linkages. A scoping study on knowledge and gaps. BRIDGE Institute of Development Studies. Universit of Sussex Brighton UK. Available from: http://www.bridge.ids.ac.uk/. (accessed on 7th August, 2011).

Buchler, S. 2009. Gender, water, and climate change in Sonora, Mexico: Implications for policies and programmes on agricultural income-generation. *Gender & Development,* 17, 51.

Carvajal-Escobar, Y. M., Garcia-Vargas and Quintero-Angel, M. 2008. Women's role in adapting to climate change and variability. *Advances in Geosciences*, 14, pp.277-280.

Central Statistical Authority (CSA). 2007. Summary and statistical report of the 2007 population and housing census: Population size by age and sex. Federal Democratic Republic of Ethiopia. Available from <http://www.csa.gov.et/pdf/cen2007_firstdraft.pdf >(Acessed on 12th August, 2011).

Daniel, R. K. 2005. Rethinking pastoralism and African development: A case study of the Horn of Africa. Available from: www.codesria.org/IMG/pdf/kandagor.pdf. (accessed on 12th August, 2011) .

Dankelman, I. 2002. Climate change: learning from gender analysis and women's experiences of organizing for sustainable development. In *Gender, development and climate change*, edited by Masika, R. Oxford: Oxfam, pp. 21-29.

Denton, F. 2002. Climate change, vulnerability,impacts and adaptation: Why does gender matter? In *Gender, development, and climate change*, edited by Masika Rachel, Oxford: Oxfam, pp.10-21.

Denton, F., Sokona, Y., Thomas, J.P. 2000. Climate change and sustainable development strategies in the making: What should West African countries expect? OECD, Environnement et Développement du Tiers Monde (ENDA-TM), Dakar, Sénégal. Available from: http://www.oecd.org/env/cc. (accessed on 12th August, 2011).

Dhanashri, B. 2010. Gender dimension of climate change adaptation: An exploration into the perceptions of women and the community. Watershade Organization Trust. Pune, India. Available from: www.wotr.org. (accessed on 3rd August, 2011).

Diro, G.T, Thomas, T. and Len, S. 2011. Ethiopian rainfall in climate models. In *African climate and climate change: Physical, social and political perispectives, Advances in global changes research,* edited by Charles, J. R. W. and Dominic, R.K.. 51-69 vol. 43. London and New York: Springer Dordrecht Heidelberg.

Fassil, K., Diress, T. & Synnevåg, G. 2001. *Traditional coping strategies of the Afar and Borana pastoralists in response to drought,* no, 17. Dryland Coordination Group (DCG).

Fiona, F. 2007. Implications for gender, pastoralism and rangeland management. In *Gender and pastoralism* Vol. 1. *Rangeland and resource management in Ethiopia*, edited by Andrew, R., Getachew, M. and Fiona, F.. Addis Ababa: SOS Sahel, pp. 71-77.

Franzel, S., Coe, R., Cooper, P., Place, F., and Scherr, S.J. 2001. Assessing the adoption potential of agroforestry practices in sub-Saharan Africa. *Agricultural Systems*, 69, 37.

Fratkin, E. M. & Roth, E. A. 2005. *As pastoralists settle: Social, health, and economic consequences of pastoral sedentarization in Marsabit District, Kenya,* New York, Kluwer Academic.

Fratkin, E., and Smith, K. 2005. Women's changing economic roles with pastoral sedentarization. In *As pastoralists settle: Social, health, and economic consequences of pastoral sedentarization in Marsabit District, Kenya, edited by* Fratkin, E. M. & Roth. E. A, New York: Kluwer Academic, pp.155-172.

Getachew, M. 2007. 'Community?' forest management in Borana. In *Gender and pastoralism: Rangeland and resource management in Ethiopia*, edited by Andrew, R., Getachew, M. and Fiona, F. vol. 1, Addis Ababa: SOS Sahel, pp.15-30.

Gurung, J. D. & Mwanundu, S. 2006. Gender and desertification: Expanding roles for women to restore drylands. International Fund for Agricultural Development (IFAD). Available from: http://www.ifad.org/pub/gender/desert/gender_desert/pdf. (accessed on 4th August, 2011).

Hassen, A. 2008. Vulnerability to drought risk and famine: Local responses and external interventions among the Afar of Ethiopia, a study on the Aghini Pastoral Community; PhD dissertation, University of Bayreuth, Germany.

Helland, J. 1980. An analysis of Afar pastoralism in the north eastern range lands of Ethiopia. Addis Ababa: International Livestock Center for Africa (ILCA).

Hesse, C. and L. Kotula. 2006. Climate change and pastoralists: Investing in people to respond to adversity, sustainable development opinion: IIED Publication. Available from: http://www.eldis.org/vfile/upload/1/document/0708/DOC23101.pdf. (accessed on 12th August, 2011).

International Panel on Climate Change (IPCC). 2007. Climate change: Impacts, adaptation and vulnerability. IPCC Working Group II Report, Chapter 19, IPCC. Available from: http://www.ipcc.ch/pdf/assessment-report/ar4/wg2/ar4-wg2-chapter19.pdf. (accessed on 10th August, 2011).

Katherine, H. 2008. *Ecology of African pastoral societies*: James Currey Oxford, Ohio University Press Athens, UNISA Press Pretoria.

Lambrou, Y. & Piana, G. 2006. Gender: The missing component of the response to climate change. Rome: FAO, Gender and Population Division Sustainable Development Department. Available from: http://www.fao.org/sd/dim_pe1/docs/pe1_051001d1_en.pdf. (accessed on 10th August, 2011).

Lemlem, A. & Yemane, B. 2007. Coping with drought in the Borana rangelands. In *Gender and pastoralism: Rangeland and resource management in Ethiopia,* edited by Andrew, R., Getachew, M. and Fiona, F. Vol.1. Addis Ababa: SOS Sahel, pp. 59-70.

Long, N. and A. Long. 1992. *Battlefields of knowledge: The interlocking of theoryand practice in social research and development.* London: Routledge.

Mannke, F. 2011. Key Themes of local adaptation to climate change: Results from mapping community-based initiatives in Africa. In *Experiences of climate change adaptation in Africa,* edited by Filho, W. Hamburg: Springer, pp. 17-32.

Markakis, John. 2004. *Pastoralism on the margin.* London: Minority Rights Group International.

Masika, R. 2002. Editorial, gender, development and climate change . In *Gender, development, and climate change.* Oxford: Oxfam, pp. l3-9.

Meze-Hausken, E. 2000. Migration caused by climate change: How vulnerable are people in dryland areas? *Mitigation and Adaptation Strategies for Global Change,* 5, 379.

Mitchell, T., Tanner, T. and Lussier, K. 2007. *We know what we need: South Asian women speak out on climate change adaptation.* London: Action Aid International and the Institute of Development Studies (IDS).

Nagar, S. E. H. E. 2001. Changing gender roles and pastoral adaptations to market opportunity in Omduruman, Sudan. In *African pastoralism: Conflict, institutions and government,* edited by Mohamed Salih, M., Ton, D. and Ahmed, A. G. M. Addis Ababa: OSSREA.

Neely, C., Bunning, S., and Wilkes, A. 2008. Review of evidence on dryland pastoral systems and climate change: Implications and opportunities for mitigation and adaptation. Land and Water Discussion Paper. Land Tenure and Management Unit (NRLA), Land and Water Division. Rome: FAO. Available from: http://www.fao.org/uploads/media/LWdisc_paper8_temp.pdf. (accessed on 10th August, 2011).

Reuveny, R. 2007. Climate change-induced migration and violent conflict. *Political Geography,* 26, pp.656-673.

Songok, C.K., Kipkorir, E.C., Mugalavai, E.M., Kwonyike, A.C. & Ng'weno, C. 2011. Improving the participation of agro-pastoralists in climate change adaptation and disaster risk reduction policy formulation: A case study from Keiyo District, Kenya. In*Experiences of climate change adaptation in Africa,* edited by Filho, W.L., Hamburg: Springer, pp.54-68.

Terry, G. 2009. No climate justice without gender justice: an overview of the issues. *Gender and Development,* 17 (1), pp.5-18.

United Nations Development Programme (UNDP). 2000. Afar pastoralists face consequences of poor rains. Rapid Assessment Mission: 19 - 24 April 2000. United Nations Development Programme Emergencies Unit for Ethiopia.

——.2008. Human development report 2007/2008. Fighting climate change: Human solidarity in a divided world. United Nations Development Programme.

——.2009a. Facts and figures on poverty. Available from: http://www.teamstoendpoverty.org/wq_pages/en/visages/chires.php. (accessed on 3rd August, 2011).

——.2009b. Resource guide on gender and climate change. Available at: www.content.undp.org/go/cms-service/download/asset/?asset_id=1854911. (accessed on 1st August, 2011).

United Nations and International Strategy for Disaster Reduction (UN-ISDR). 2008. Gender perspectives: Integrating disaster risk reduction into climate change adaptation: Good practices and lessons learned. Available from: www.unisdr.orgGender_Perspectives_Integrating/Gender_Perspectives_Integrating_DRR_CC_Good%20Practices.pdf. (accessed on 8th August, 2011).

Valerie, N., Kate, M., Terry, C., John, M. & Adrienne, M. 2002. Uncertain predictions, invisible impacts, and the need to mainstream gender in climate change adaptations. In *Gender, development, and climate change*, edited by Rachel, M. Oxford: OXFAM, pp.51-59.

Wamukonya, N. & Skutsch, M. 2002. Gender angle to the climate change negotiations. *Energy & Environment*, 13 (1), pp.115-124.

Wawire, Violet. K. 2000. Gender and the social and economic impact of drought on the residents of Turkana District Kenya. *Gender Issues Report Series*, 21. Addis Ababa: Organization for Social Sciences Research in East Africa (OSSREA).

Welde-Georgis, T. 1997. El Niño and drought early warning in Ethiopia. University of Colorado at Boulder - Consortium for Capacity Building (CCB), INSTAAR . *Online Journal of African Studies*, No. 2. Available from: http://papers.ssrn.com/sol3/papers.cfm?abstract_id=1589710. (accessed on 14th May, 2011).

Yemane, B. 2004. *Resource shrinkage and possible causes of conflict in the Afar National Regional State, gender, resource and conflict management.* In *Research and development experience on dryland husbandry in Ethiopia,* edited by M Haile, M. Wolde-Georgise, D. and Teka, T. Dryland Husbandry Project (DHP)Ethiopia: OSSREA, and Mekelle Unviersity.

Chapter 5

Effects of Climate Change and Variability on Pastoralist Women's Accessibility to Social Services: Case of Fentalle Woreda, Central Rift Valley of Ethiopia

Getnet Feyissa[*]

Abstract

Effects of climate change and variability continue to knock on the doors of each society particularly that of the agrarian and pastoral. Women in pastoral communities are more vulnerable and prime victims as the environment turns to be increasingly worse. This is a research conducted to assess the climatic and spatial phenomenon that exacerbates the already poor living conditions of the pastoral women in Fentalle *Woreda* (district) of central Ethiopia. The methodologies in the research include the conceptual frameworks and empirical formulae and software analysis in the ArcGIS environment using Multi-Criteria Evaluation. The temperature of Fentale *Woreda* has shown a 0.47°C increase per decade as its total rainfall has remained constant for the years 1972 - 2010 with high inter-annual variability. Pastoral women's access to social services faces severe challenges due to climate change and variability which is being aggravated by multiple variables and spatial factors: proximity to water source, school and health centres, as well as slope characteristics of the locality.

[*]getfiran@gmail.com

Introduction

Background to the Study

Pastoralism is a finely-honed symbiotic relationship between local ecology, domesticated livestock and people in resource-scarce, climatically marginal and highly variable conditions (World Initiative for Sustainable Pastoralism 2010). It represents a complex form of natural resource management; involving a continuous ecological balance between pastures, livestock and people (Nori and Davies 2007). Among pastoral communities, climate variability and change, such as higher temperatures and irregular rainfall, reduce pasture and water, leading to animal loss through hunger and disease (Jotoafrika 2011). The pastoral areas of Ethiopia are faced with frequent natural disasters, many of them weather-related. However, the deterioration of the natural environment due to unchecked human activities and poverty has further exacerbated the situation (Abebe 2007). Ethiopia is especially highly vulnerable to drought, making drought the single most important climate-related natural hazard impacting the country from time to time (Ministry of Agriculture 2011). Drought and the delayed onset of the rainy season lead to poor grass regeneration, forage scarcity, water shortage and heat stress on livestock. This has, in turn, increased the mortality rate of the livestock, vulnerability to diseases and physical deterioration due to long distance travel for water and pastures (Muluneh and Demeke 2011). Although drought may occur anywhere in the world, its damage is not as severe as in Africa, in general, and in Ethiopia, in particular (Jotoafrika 2011).

It is important to keep gender in mind when examining the issue of climate change because gender inequalities intersect with climate risks and vulnerabilities (Aguilar 2009; BNRCC n.d). For example, with increasing drought, it is women who have to walk longer distances to collect water. Women are often the main repositories of vital local and traditional knowledge (LEG 2002; UNDP 2009). There is a causal inter-relationship between climate change and gender:

- Climate change tends to exacerbate existing gender inequalities.
- Gender inequalities lead women to face larger negative impacts.

The vulnerability of women to the effects of climate change is largely due to their dependence on natural resources, their responsibility for water and food procurement and their increased risk exposure during times of disasters and severe weather crises (UNDP 2008).

Statement of the Problem

The effects of climate change and variation are worse for pastoralists who live in under-developed and remote areas. A review of the literature shows that the studies conducted so far on how climate change along with other spatial factors affect pastoral women is not yet well documented. Specifically, no research has analysed climate change and other spatial factors as the main impediment for Ethiopian pastoral women's access to basic social services, including schools, water sources, grasslands and health posts. In Fentalle *Woreda* where pastoralism is the dominant economic activity, the effect of climate change is worse as vividly observed by a recurrent dependence of people on direct relief food aid; and the sheer number and the presence of safety net programmes in the district compared to the neighbouring *woredas*. This is supported by information obtained from East Shewa Disaster Prevention and Preparedness Office.

This study is designed to fill existing gaps in previous studies by bringing spatial points of view and creating a new way for mapping vulnerability of women using multi-criteria analysis of GIS applications which over-lays the layers of various spatial phenomenon's and variables. The resultant climate vulnerability map can serve governmental and non-governmental organisations, local planners and scientists as a primary source of information and a tool for policy-making.

Objectives of the Study

Overall, the objective of the study is to assess and map spatial phenomenon which exacerbate the already poor living conditions of pastoral women of the *woreda* due to climate change and variability and to prepare pastoralist women's vulnerability index to climate change and variability at *kebele* (lowest administrative unit under *woredas* or sub-city administrations) and *woreda* levels. The specific objectives are to:

- identify and analyse the multiple drivers of climate change and vulnerability in the study area;
- map the spatial variables which have direct and indirect impact

on pastoral women of the *woreda* caused by climate change;

- identify and analyse factors causing and exacerbating both bio-physical and social vulnerability of women pastoralists; and
- prepare pastoralist women's vulnerability index to climate change for the *woreda*.

By focusing on the above objectives, the study seeks to answer the following research questions:

- How does climate change and variability affect pastoralist women's access to social services?
- What are the trends of major climatic elements and their rate of change in the study area?
- What are the spatial factors that exacerbate the situation of climate change and vulnerability?
- What are the local adaptive coping strategies undertaken by pastoralist women when extreme situations occur due to climate change and variability?

Methodology

The study area is found in the Central Rift Valley of Ethiopia in Oromia Region, between 8^{0}39'55"N' - 9^{0}9'15"N latitude and 39^{0}39'19"N - 40^{0}5'46"E longitude with a total area of 1533 km^2 (see Figure 1). The area has arid and semi-arid climate, with low and erratic annual rainfall, low altitudes and warm temperatures. The area enjoys long rainy season locally known as *Ganna*, lasting from June to August, and short dry season called *Birra* from September to November. The Kerreyu are the indigenous inhabitants of the Metehara Plain and Mount Fentale area (Ayalew 2009; Shimelis and Dafa 2009). According to the 2007 census results, the population density of the district was 77 persons per km^2 (Getnet 2011).

Figure 1: **Map of the study area**

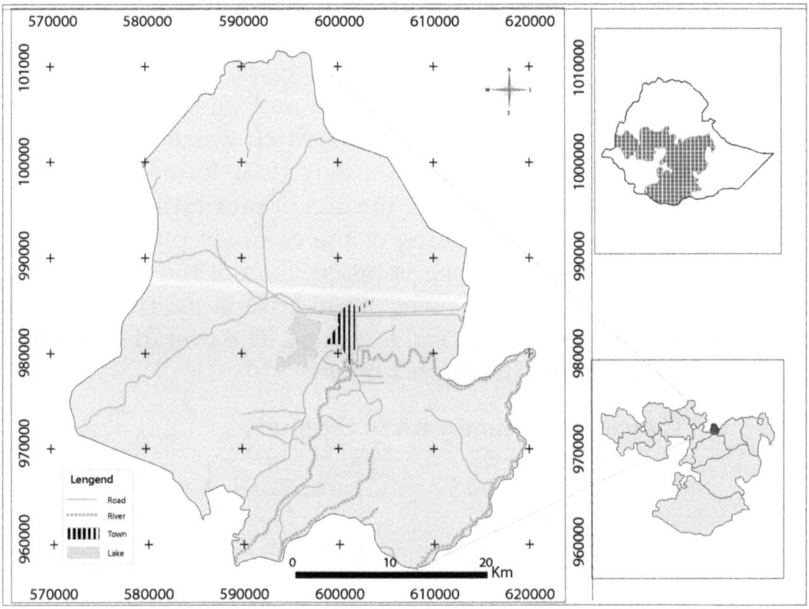

Source: Central Statistical Agency of Ethiopia, map prepared by the Author
Vertical Axis: Northing Value
Horizontal Axis: Easting Value

Both primary and secondary data sources were used. Five data sources were identified, namely: water source, school points, health centre, locality points for women population density and meteorological station points. Data were collected through field survey using portable GPS. Secondary data on socio-economic characteristics were obtained from the *woreda* office and reports compiled by the Central Statistical Agency while meteorological data were gathered from the National Meteorological Service Agency (NMSA) of Ethiopia. Satellite imagery and Shuttle Radar Topographic Mission (SRTM) data were obtained from United States Geological Survey (USGS). A questionnaire was used to generate primary information.

Female pastoralists were randomly selected based on the strata of administrative boundaries to fill out the questionnaire (see Annex 1 and Table 3). The total sample size of the study was 300 female pastoralists. All spatial and relevant data were analysed using ArcGIS 9.3 and E RIDAS Imagine 9.2 software. The multi-criteria variables are: distance from schools, distance from health centres, and from water source and slope class. Indices generation considered two perspectives for analysis and integration: the first one is the multi-criteria analysis of GIS which was used to analyse geo-referenced data. The second is empirical formulae, used to calculate vulnerability index with the aim of integrating spatial and social variables for the modelling of the complex phenomenon of the problem. The resultant map is the product of the vulnerability index displayed graphically using multi-criteria analysis and the values derived from the empirical formulae. The overall conceptual framework is presented in Figure 2 below.

Figure 2: **Conceptual framework of the study**

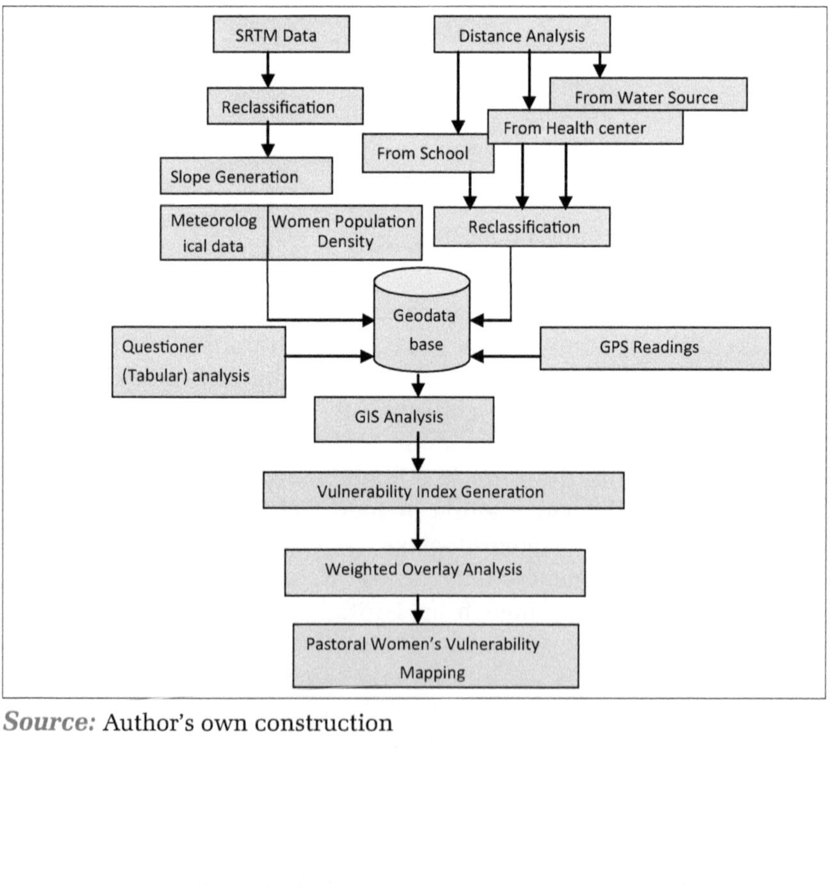

Source: Author's own construction

Theoretical and Conceptual Literature

Climate Change and Variability and Vulnerability

According to United Nations Framework Convention on Climate Change (2006), climate change is an adjustment of climate caused directly or indirectly by human activity that alters the composition of the global atmosphere and by natural variability observed over comparable time scales. Climate variability, on the other hand, is the departure from the normal or the difference in magnitude between climatic occurrences (Kinyangi *et al.* 2009). The impact of climate change depends on hazard and vulnerability. Hazard is the susceptibility to the physical effects of climate change, for example, drought, flood, storms and changes in weather patterns; while vulnerability is the sensitivity of a country's physical, economic, social and political systems to the effects of climate change, including its ability to anticipate and adapt to those effects (UNICEF 2008). In line with this, IPCC (1995) in its Second Assessment Report defines vulnerability as the extent to which climate change may damage or harm a system. It adds that vulnerability depends not only on a system's sensitivity, but also on its ability to adapt to new climatic conditions. The IPCC's (2001) definition on a region's vulnerability to climate variability and change is described by three elements: exposure, sensitivity and adaptive capacity. While climate change is a global phenomenon, its negative impacts are more severely felt by poor people in developing countries who rely heavily on the natural resource base for their livelihoods. In particular, poor rural communities relying greatly for their survival on agriculture and livestock keeping, which are amongst the most climate-sensitive economic sectors (IFAD 2009), are most affected. This view was also shared by the Inter-governmental Panel on Climate Change (2007) where it affirms that climate change is real; will become worse; and that the poorest and most vulnerable people will be the worst affected. The more vulnerable members of a household (women, children and the elderly) are, therefore, exposed to food security risks in a difficult environment and often they have no residual strategies for survival other than to access relief from aid agencies (Kinyangi *et al.* 2009).

Women's Vulnerability to Climate Change

Climate-change vulnerability indices combine a variety of economic, social, environmental and political data to estimate countries' or regions' likely risk of climate damages (Stanton et al. 2011).Women, particularly in least-developed countries (LDCs), are disproportionately vulnerable because of their dependence on bio-fuels and other natural resources; responsibility for water procurement and household care; role in securing food and fuel; pre-dominant presence in low technology agriculture, and greater exposure to risk in crisis and severe weather events that may have been influenced or impacted by climate change (UNDP 2009). It should be noted that although women are especially vulnerable to climate change and its impacts, at the same time they have special capacity as agents of change in the face of climate change: they manage many of the world's agricultural resources and are also likely to have primary responsibility for raising children UNDP 2009). However, all over East Africa, livestock ownership and use rights are varied and complex and not similar. In some places, men own complete exchange rights over all stock and at other places, they control only larger stock (Olson 2010). But what makes women's case different from men's is that, female-headed households who are left behind when men migrate as a result of extreme occurrences and disasters are often the poorest. As wives of seasonal migrants, they may not receive remittances or have alternative sources of income. The workloads of these women, their children and the elderly, increase significantly as a result of male out-migration, which is likely to expand as climate change-related impacts escalate. Additionally, where there is increased male out-migration due to resource shortages and unreliable job markets caused by climate change, women are left behind with additional agricultural and household duties (UNDP 2009).

GIS Multi-Criteria Analysis and Climate Change and Vulnerability Mapping

Multi-criteria evaluations represent the means for an integrated and participative approach, capable of considering the perspectives of all the agents and the behaviour of all the environmental components playing a part in territorial processes. Despite most spatial decisions, problems are multi-criteria in nature; its techniques have only recently been inserted and integrated in spatial analysis, attaining the concept of multi-criteria spatial decision systems (Lapucci, Lombardo and Petri 2005). The GIS analysis and integration with statistical data or multi-criteria could make a GIS application more flexible and accurate (Cavallo and Norese 1998; Mokarram and Aminzadeh 2010). In view of the above literature, the multi-criteria analysis (MCA) is a method of analysis that considers separately a variety of different criteria for assessing a project and then attempts to define the trade-offs between the criteria in order to decide on the best course of action. Preferably, the analysis would also show various alternatives for comparison (Abebe 2007). Scholars agree that a Spatial Decision Support System (SDSS) usually relies on a GIS that has the additional capability of dynamically displaying any query, report or piece of information in the system as a map. Both SDSS and GIS use geo-spatial technology and the difference from other types of information systems is their capability to display information. In a GIS, almost any type of data can be stored, analysed spatially and modelled across time (Airine and Homocianu 2009). Currently, Multi-Criteria Decision-Making and GIS integration has resulted in various query solutions, ranging from maps based on non-compensatory decision rules and crisp feature visualisation to geographical spaces integrated with sophisticated compensatory decision techniques (Malczewski 1999). As climate change is a spatial phenomenon and pastoralism is practised in a certain geographic area consisting of a population women live in, assessing the situation using GIS Multi-Criteria is invaluable.

Definition of Model Variables and Vulnerability Index Specification Indices

In the literature, quantitative assessment of vulnerability is usually done by constructing a 'vulnerability index'. This index is based on several sets of indicators that result in vulnerability of a region. It produces a single number which can be used to compare different regions. Literature on index number construction specifies that there should be good internal correlation between these indicators. IPCC's (2001) definition on region's vulnerability to climate variability and change is described by three elements of exposure, sensitivity, and adaptive capacity, as follows:

Exposure: This is the nature and degree to which a system is exposed to significant climatic variations. The exposure of a system to climate stimuli depends on the level of global climate change, and due to the spatial heterogeneity of anthropogenic climate change, on the system's location.

Sensitivity: This is the degree to which a system is affected, either adversely or beneficially by climate-related stimuli. The effect may be direct or indirect.

Adaptive capacity represents the potential to implement adaptation measures that help avert potential impacts.The relevance of this criterion depends on the relationship between the indicators and the construct they are supposed to measure (ICRISAT 2009).

Vulnerability may be formulated mathematically as follows:

$$V = f\,(I - AC)$$

(-) (+)

Where V is vulnerability, I is potential impact, and AC is adaptive capacity.

Adaptive Capacity: A higher adaptive capacity is associated with a lower vulnerability, while a higher impact is associated with a higher vulnerability. Climate-change vulnerability indices combine a variety of economic, social, environmental and political data to estimate countries' or regions' likely risk of climate damages. Some include a wide array of indicators and serve as holistic depictions of a multitude of climate-change risks, while others are designed to gauge a single threat. Most share common elements that provide the means to examine key climate vulnerabilities from two important perspectives: adaptive capacity, and measures of sensitivity to damages. A few vulnerability indices layer the effects of globalisation on top of climate change to reveal doubly exposed hotspots of vulnerability (Stanton *et al.* 2011). In this case, the variables are selected based on the indicator method since indicator methods are based on selecting some indicators from the whole set of potential indicators and then systematically combining the selected indicators to indicate the levels of vulnerability. These levels of vulnerability may be analysed at local, national, regional and global scales (Temesgen, Hassan and Claudia 2008).

Data Presentation and Analysis

Based on existing literature ((ICSSR 1983; IPCC 2007; UNDP 2009; Getnet 2011), any climate change and variability is indicated or studied by analysing the trends of major climatic elements.

Temperature and precipitation are the main elements available in most meteorological stations. In the case of Ethiopia, specifically in Fentalle *Woreda*, the two elements with their derivatives have been used. The monthly average maximum and minimum temperatures and total annual rainfall of the woreda for 39 years have been considered for this research.

Trends in Precipitation

Its lower topography and location give Fentalle *Woreda* warm climatic conditions. The 39 years rainfall record taken from Metehara Sugar Factory shows a great inter-annual variation both in intensity and duration. The average total annual rainfall for the years 1972 - 1990 is 536.7mm. The average annual total rainfall from 1991 to 2010 is 559.1mm. Hence, the average annual total for the last 39 years is 548 mm (see Figure 3 below).

Figure 3: **Trends in total annual rainfall: 1972 - 2010)**

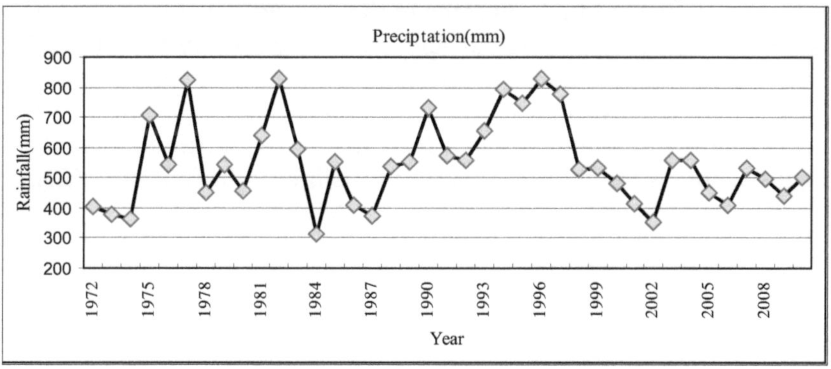

Source: Author's own construction, with data from National Meteorological Service Agency of Ethiopia

The frequency of occurrences of extreme high rainfall record has decreased since 1997. However, this does not mean that its intensity has also been decreasing. The frequency of minimal total annual rainfall has been increasing in recent years. Before 1997 the total rainfall above 700mm had been recorded eight times, even though after 1997, the annual rainfall over 700mm has not been recorded. The number of years with total minimum rainfall has shown increment as indicated in a rainfall anomaly of Figure 4.

Figure 4: **Precipitation anomalies, by year**

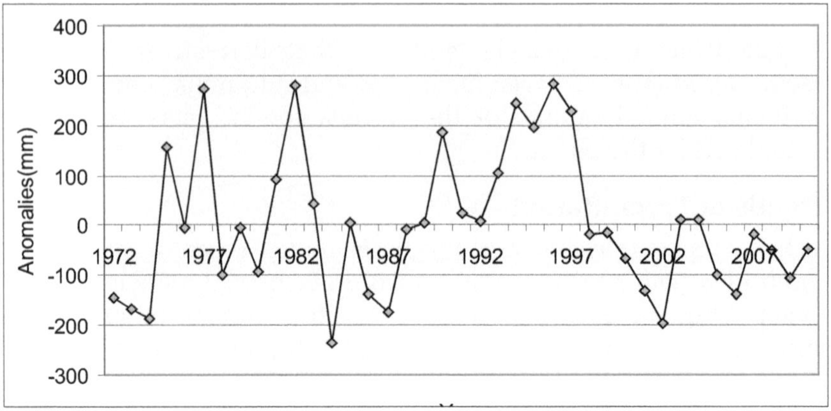

Source: Author's own construction, with data from National Meteorological Service Agency of Ethiopia

The other important point is the monthly variation of rainfall. June, July and August are rainy months in the *woreda*. The highest rainfall is recorded in July. November and December are dry months, with nearly no rainfall recorded. The highest coefficient of variation is 26 per cent. Due to this climate variability, the pastoralists in the *woreda* are compelled to move long distances during dry seasons. Figure 5 shows the monthly rainfall distribution in Fentalle *Woreda*.

Figure 5: **Distribution of monthly rainfall in the study area: 1972 - 2010**

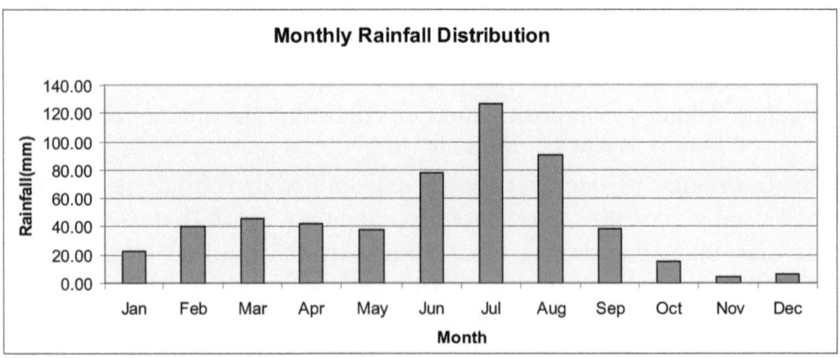

Source: Author's own construction with data from National Meteorological Service Agency of Ethiopia

The northern pastoralist *kebeles* of the *woreda* receive maximum rainfall compared to the *kebeles* in the southern parts of the *woreda*. *Kebeles* located on the highest elevation receive up to 800mm total annual rainfall while the majority of the *kebeles* in the southern parts of the *woreda* receive less than 500mm total annual rainfall as depicted in the Figure 6B. The rainfall variance within the *woreda* is proportional to the rainfall distribution and intensity. The higher the rainfall, the lower the variance and vice versa (see Figure 6A).

Figure 6A: **Rainfall variance N-score**

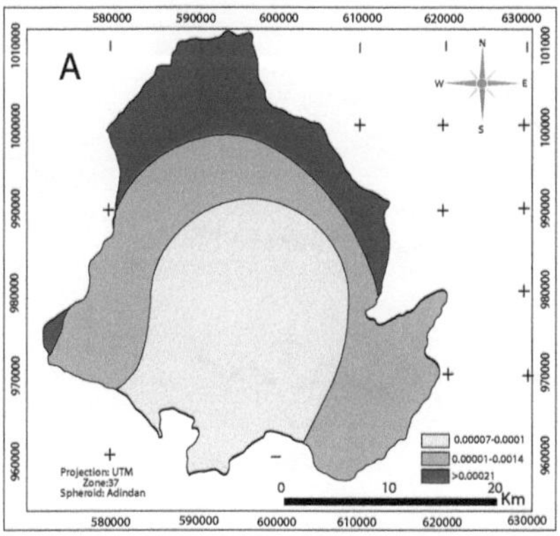

Figure 6B: **Rainfall distribution**

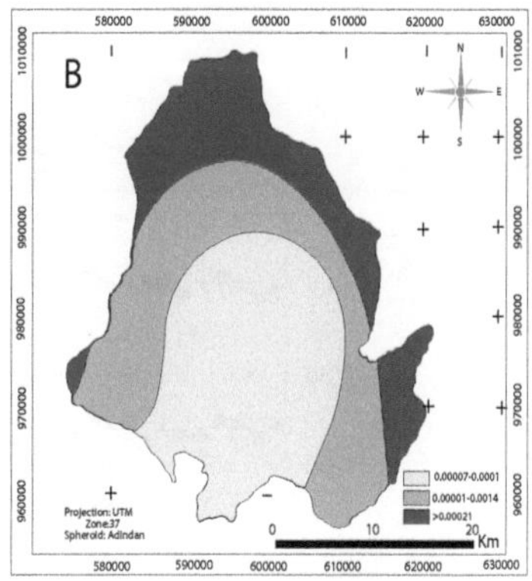

Source: Author's own construction, data from Meteorological Service Agency of Ethiopia

Temperature Analysis

Considering the decadal increment of mean temperature, the rate at which temperature increases varies, Fentalle *Woreda* has both hot and warm climatic conditions. Compared to the 1972 - 1980 period, the minimum temperature in the 1981 – 1990 period increased by 0.08° C. However, the rate of change in minimum temperature in the 1991 – 2000 period has increased by 0.75°C compared to the 1981 - 1990 decade. For the recent decade of 2001 - 2010, the rate of change for minimum temperature was at 0.32°C in contrast to the preceding decade. But the maximum temperature increased at a rate smaller than the minimum temperature. As can be seen from Figure 7, the maximum temperature for Fentalle *Woreda* increased only by 0.3°C in between 1972 - 1990 while between 2000 - 2010 it rose by 0.99 °C. As mean temperature is the average of minimum and maximum temperature, it also increased by 0.47°C per decade for the years 1972 - 2010. Between the years 1972 - 1980 and 1981 - 1991 the rate of temperature increased by 0.06°C while it increased by 0.62°C and 0.73°C in the decades between 1991 - 2000 and 2000 – 2010, respectively. Therefore, the overall average mean temperature for 1972 - 2010 was 25.17°C.

Figure 7: **The trend in maximum, minimum and mean temperature for Fentalle *Woreda*, by years**

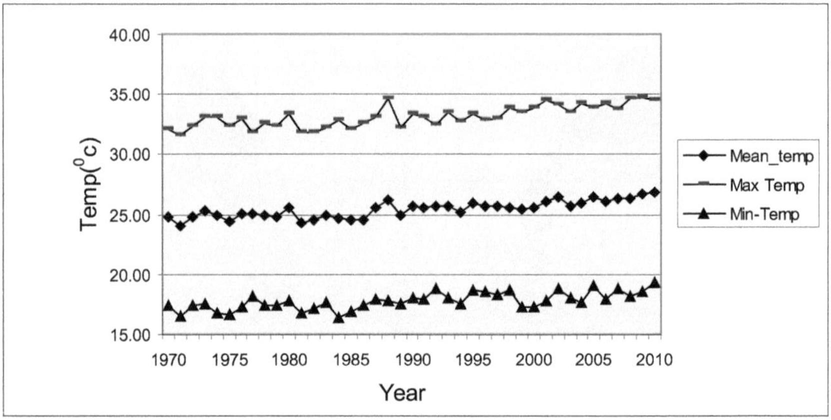

Source: Author's own construction, with data from National Meteorological Service Agency of Ethiopia

Geographically, as it is interpolated from the neighbouring meteorological stations along with Metehara Sugar Station, central parts of the *woreda* have a maximum temperature of more than 26°C as indicated on Figure 7. The normalised temperature variance shows a slight spatial difference, since the topography of the study area does not have a marked topographic difference. Between 1972 and 1990, the average range of temperature was 15.27°C while its value rose to 15.46°C for the years 1991 - 2010. Figures 8A and 8B indicate the temperature distribution and N-score variance of temperature in the *woreda*.

Figure 8A: **Mean temperature**

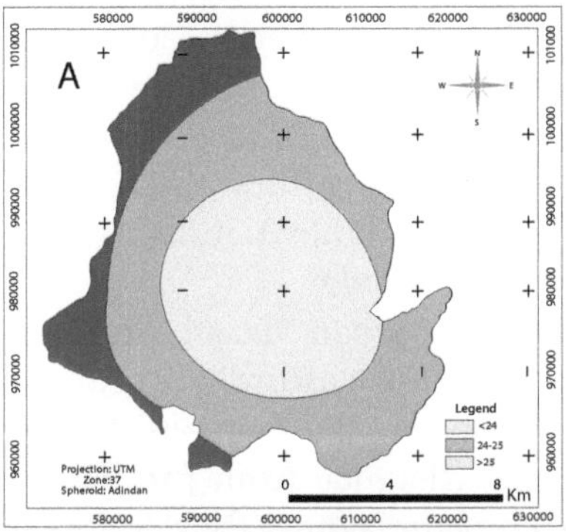

Figure 8B: **N-score temperature variance**

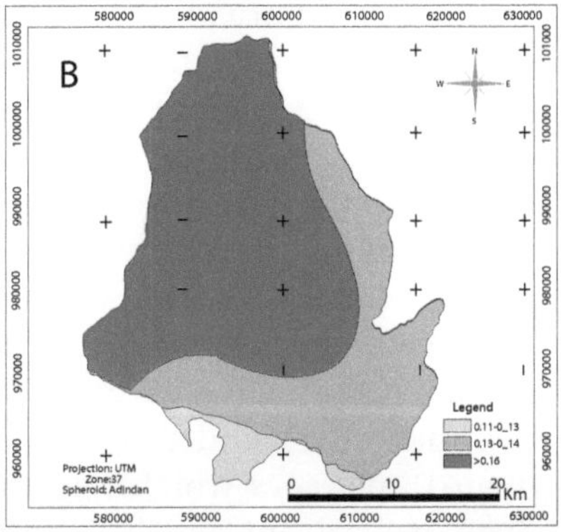

Source: Author's own construction, data from National Meteorological Service Agency of Ethiopia

Distance Analysis

The problem of distance on accessing basic social services to pastoralist women is analysed using a questionnaire administered to 300 pastoralist women. In order to consider the potential impacts created under climate change on pastoralist women, here three straight line distances are analysed which this research considers as a basic phenomenon for spatial distribution of pastoralist women and considerations along with the data generated by using a questionnaire. Pastoralism is highly related to the spatial movements which have direct and indirect impacts along with the topography of the area. These are distances from water source, distances from school, and distances from health centres. For all data, the unit of measurement is in kilometres. All the values use the kilometre as a unit of measurement and ArcGIS 9.3 for the analysis. The spatial analysis methods applied in all rasterisation of the values of point features are inverse distance interpolation.

Distance from Water Source

The water sources used for the analysis in the study area are rivers, lakes and water points (major borehole points). The water layer is, therefore, the cumulative of the three sources. Beseka is the only lake layer used for the analysis. The lake has relatively slight usage for the pastoralists since it has salty water. Regarding rivers, Dofa Mukye and Wanga rivers are the intermittent streams while Kessem, Arba, Awash, Adembuty and Huku are the permanent rivers used for the analysis. These give continuous water supply service for the public as a whole. As evident from their shape and length (see Figure 9), the distance analysis was first done by over-laying the water sources and converting them to continuous surfaces. Table 1 indicates the permanent and intermittent streams in the *woreda*.

The weights of the water sources are given based on their coverage and the importance they give for the pastoralists and the influence they have in water layers based on the availability, the area they cover and usage. Accordingly, distance from the lake is given 15 per cent weight; distance from water points (boreholes) is given 35 per cent weight; and distance from rivers is given 50 per cent weight, guided by the water situation of the pastoralists. The straight line distance from one water point to the next water point is indicated in kilometres and converted to continuous surfaces.

Table 1: **Rivers in Fentalle** *Woreda*

Name of River	Type of River	Length in Km
Arba river	permanent	28.6
Awash	permanent	57.8
Dofa	intermittent	0.83
Huku	permanent	0.29
Kesem	permanent	14.9
Mukye	intermittent	0.54
Wanga	intermittent	8.5
Adembuty	permanent	0.46

Source: Central Statistical Agency of Ethiopia

The rivers have given the largest value as livestock herding is mainly dependent on them for water to drink. The women in the *kebeles* within the central part of the *woreda* which have access to all the water sources have less distances to travel to reach the water source from their place of residence measured from all direction than the pastoralist women found in the periphery of the *woreda*. The average distance from one water point to the other is 10 km, as given in Figure 9.

Figure 9: Re-classified straight line distance from water source (in km)

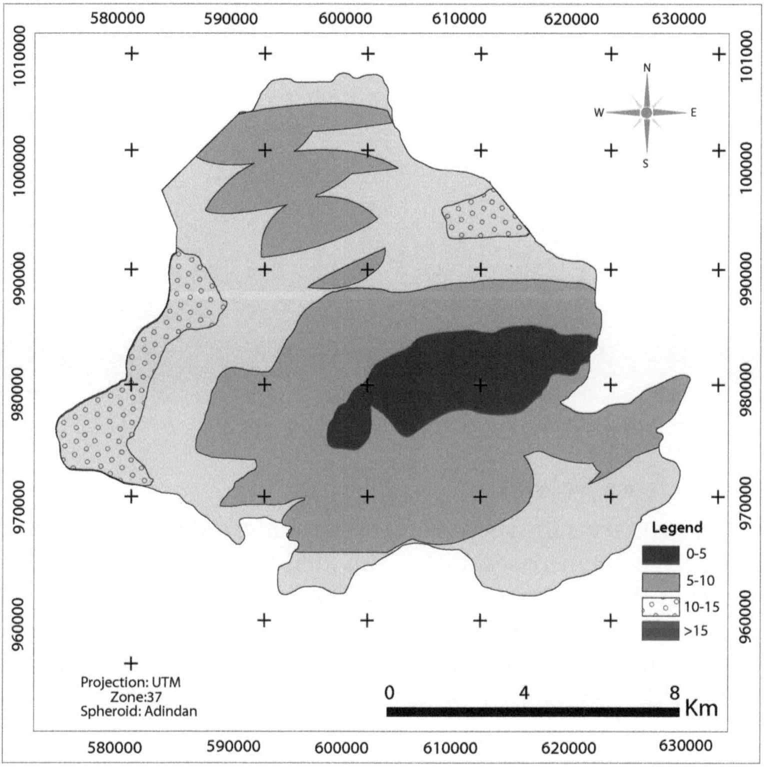

Source: Central Statistical Agency, Map prepared by the Author

Lake Beseka is the largest water body found in the *woreda*. However, since it is very salty, it is not used for drinking purposes.Research shows that the lake is increasing in size from time to time and it is invading the rangeland found in the area (Figure 10).

Figure 10: **Partial view of Lake Beseka**

Source: Photo taken by the Author, on March 2010.

Distance from School

Nine (9) primary schools were used to analyse the school distance covered by women pastoralists. The analysis was premised on the notion that the distance from school would have an impact on the women's adaptation mechanism and awareness about the changing environment. It relies on the fact that as women goes far to get assess to school, the probability of school dropout will be high, and the awareness and coping strategy will be low, as compared to women travel short distance to get school since the probability of school dropout is to the minimum. The study considered only primary schools. The data show that the shortest average distance from one school to the other is 3 km while the longest distance travelled by women is 15 km. The schools are taken as point features by hand-held GPS. Based on their distribution and location of residence, and infrastructures and topography of the area, the central *kebeles* have the shortest distance to the school while others have to travel as far as 15 km to get primary education which benefits indirectly the pastoralist women and their children as the time spent to travel to school will be saved and the children help their parents in the pastoralist community. On average the nearest primary school distance was 8 km. Figure 11A presents a proximity map of the school.

Re-classified Distance to Health Centres

Health centres within the *woreda* have been identified as point features and converted to continuous surfaces using inverse weighted distance interpolation method. The analysis rules considered here are similar to the previous layer (school). It was assumed that the less the straight line distances between the health centres, the more the vulnerability of women pastoralists to climate change-related diseases. The proximity map of the health centres is given in Figure 11B. As evident from the figure, a woman has to travel on average 14 km to reach a health centre. However, women in the remote parts of the *woreda* (northern *kebeles*) have to travel up to 28 km to get a health centre facility.

Figure 11A: **Re-classified distance from school**

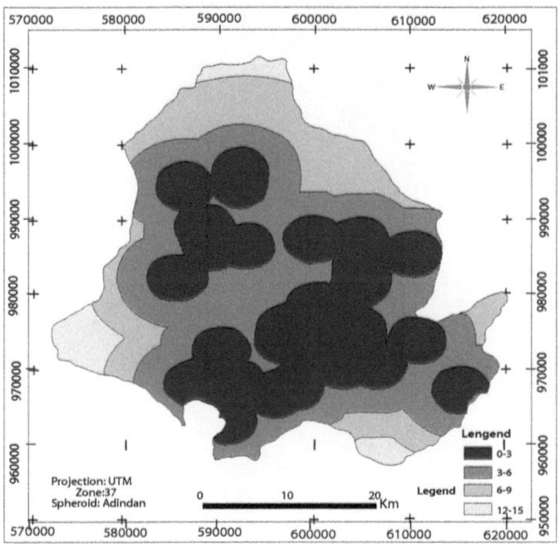

Figure 11B: **Re-classified distance from health centre**

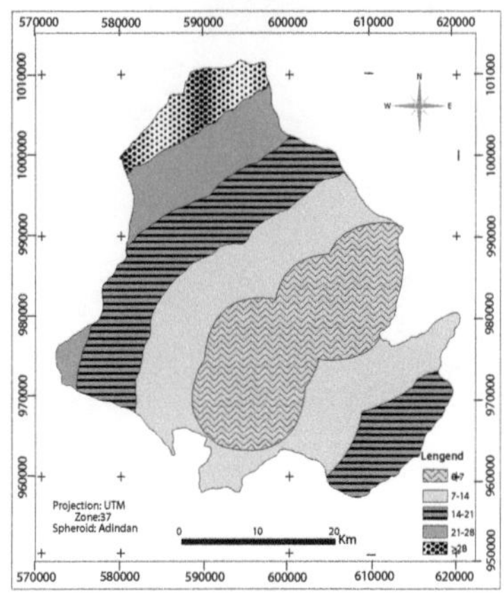

Source: Author's own construction, data from Central Statistical Agency of Ethiopia

Re-classified Slope

Based on the digital soil and terrain database of East Africa notes on the arc/info files released in version 1.0 in April 1997, the slope classes for the terrain of East Africa is divided into three classes. These are given as lowx(0%-8%) Highx (8% - 30%) and SlopePlusx (above 30%)

Based on this standard, the slope class of the Fentale *Woreda* is divided into four classes: 0^0 - 8^0, 8°- 16^0, 16^0 - 30^0 and above 30^0. Except for the main asphalt road which crosses the *woreda*, almost all pastoralists walk on foot or use animals for transportation. The capacity of animal transportation, however, would be affected by the slope of the area. As a rule, it is obvious that, the gentler the slope, the less the vulnerability. Figure 12 shows the slope distribution degrees. From the figure, it is evident that majority of the topography is plain and falls within the slope range of 0^0 to 8^0 and the highest slope is registered in steep side of Mount Fentale, and the value measures above 16^0.

Figure 12: **Slope map (in degrees)**

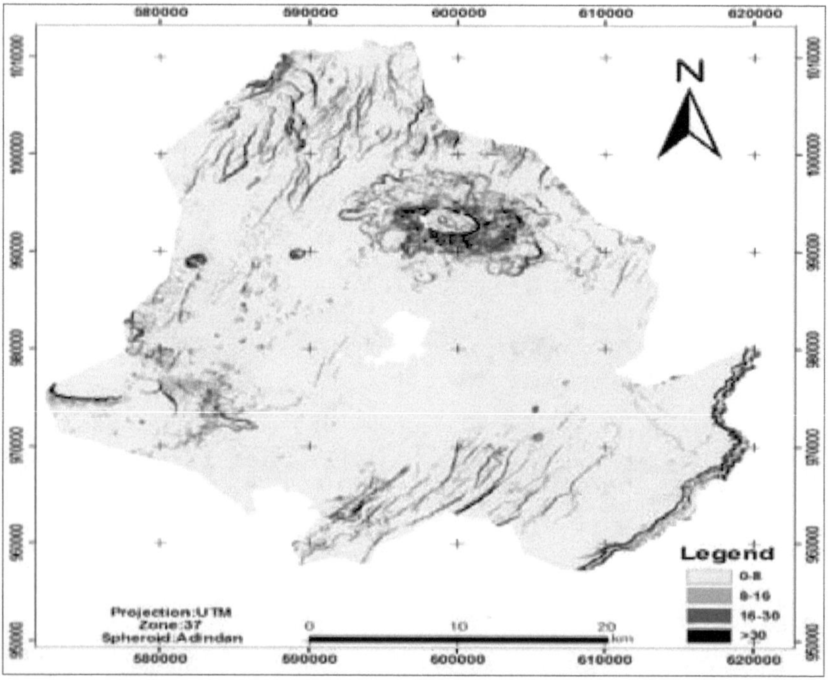

Source: Author's own construction, SRTM (30m) data from USGS

Women Population Spatial Distribution

The women population density per *kebele* plays a significant role for the result of vulnerability map; the more the population density the higher the vulnerability will be. Climate vulnerability will be manifested in the high density of population. The pastoralist women population on *kebele* level was obtained from 2007 Census of the Central Statistics Authority of Ethiopia. The female population density map was prepared by the reperesentaion as follows. In

Figure 13, one dot represents 150 households. This is premised on the standard that in rural Ethiopia in one household there are 7 persons. Among pastoralists, the number of women per household is always less than men. Therefore, the research took 3 women per household. If one dot represented 150 households it means that one dot represents 450 women. The areas where the dots agglomerate are the places where high women pastoralist population are found excluding the town women population of the *woreda*.

Figure 13: **Women pastoralist population density map**

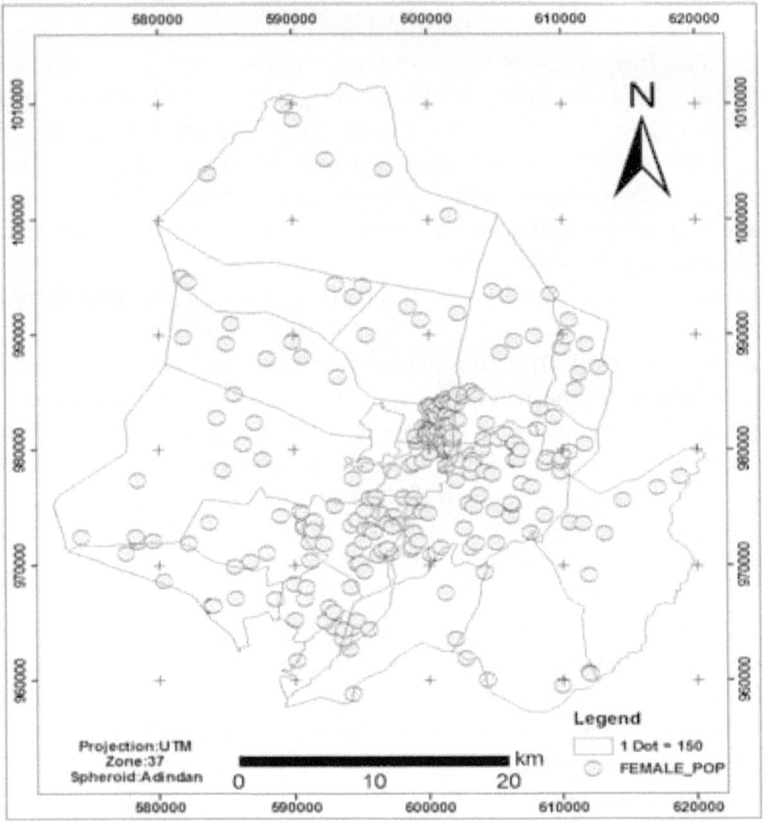

Vertical Axis: Northing Value
Horizontal Axis: Easting Value
Source: Central Statistical Authority of Ethiopia (2007)

Discussion of Findings

The climatic change and variability in Fentalle *Woreda* is increasing from time to time. The *woreda* is increasingly becoming hotter and receiving scanty annual rainfall. This climatic change has resulted in shortages of grass and water at nearby distances of the *woreda*, forcing the residents, particularly pastoral women, to travel further long distances in search of pasture and water.

The rainfall records of the *woreda* for the past 39 years show greater inter-annual variations. The average annual total rainfall since 1997 has been recorded at 567 mm, while the average annual total rainfall of the *woreda* for the last 39 years is reported to be 548mm. The frequency of the occurrences of extreme high rainfall record has declined since 1997 although the precipitation has not decreased in intensity on average. The frequency of minimal total annual rainfall has also been increasing in recent years. The question is: Why have pastoralists of the *woreda* suffered from shortage of pasture and water when there has been no decrease in rainfall intensity for long time?

The answer to this question can be given and analysed from two perspectives:

The first perspective is inconsistency in the duration, beginning and end of the rainy season. Delay or early coming of rain has become a common phenomenon in the *woreda* for many years. Information gathered through the questionnaire indicates that rain has started to stop early, leaving grasses for the livestock not fully-grown. This, in turn, results in the shortage of pasture and water for their livestock forcing the pastoralists of the *woreda* to stay for a prolonged time until the next rainy season comes or to move long distances in search of grass and water. Rainfall exceeding 700mm has not been recorded for the past 13 years. The coefficient of variation of rainfall is also the highest, standing at 26 per cent.

The second perspective is attributable to increase in temperature. Considering the decadal increment of mean temperature, the rate at which temperature increases is rising.

Based on the data gathered through a questionnaire, drought accounts for 70 per cent as livestock's diseases and income shocks account for 15 per cent for each of the challenges faced by people in the *woreda*.

On average, for each extreme weather conditions, occurred in between 1972-2010, each house hold has lost three cattle and a camel. About 70 per cent of the respondents reported that they move to better grass fields and water sites covering a distance of 50 km to escape from the harsh climate during the prolonged dry season.

The major source of income of the pastoral women in the *woreda* is the sale of livestock products. However, during the dry season this income decreases by half. Under such harsh environment, the women collect wood and produce charcoal to augment their income as a coping strategy to withstand the vulnerability of climate change.

Regarding access to markets, the majority (71.45 per cent) of the respondents reported that access to market is not a problem for the pastoral communities of the *woreda*.

The results of all the previously analysed spatial and non-spatial variables are normalised as vulnerability index. The final climate change map is produced based on the vulnerabilty index. The climate change vulnerability indexes for each pastoralist kebele are indicated on Table 2. The final and the degree of vulnerability will be given based on the normalised values given for each kebele. The vulnerability index is prepared based mainly on the formative model that all indicators have negatively contributed to the construct. These are temperature variance, rainfall variance, distance from water sources, distance from schools, distance from health centres, pastoralist women's density, and slope characteristics of the locality. Hence, vulnerability index is a formative measurement and the indicators chosen do not have internal correlation.

Table 2: **Vulnerability index**

Pastoralist Kebeles	**Normalised Scores**							
	(1)*	(2)	(3)	(4)	(5)	(6)	(7)	(8)
Gola	0.171	0.000085	0	0	0	0	1	1.171085
Haro kersa	0.171	0.00021	0.5	0.5	0.77	0.39673	0	2.341525
Ledi Fate	0.135	0.000165	0.5	0.2	0.4	0.19837	0.125	1.554761
Kobo	0.1525	0.000103	0.7	0.25	0.13	0.39673	0.125	1.756408
Dire Sadan	0.135	0.000085	0.5	0	0	0	0.583333	1.218418
Kenifa	0.135	0.000165	0.75	0.375	0.4	0.19837	0.069444	1.924205
Gelcha	0.171	0.000103	0.04	0.08	0	0	0.333333	0.624436
Gidara	0.125	0.000103	0.5	0.375	0.4	0.04891	0.208333	1.653572
Gara Dima	0.1275	0.000085	0.5	0	0.18	0.39673	0.583333	1.783749
Gederena kube	0.135	0.000085	0.66	0	0	0.39673	1	2.191815
Sarana Weba		0.000085	0.44	0.125	0.13	0.39673	0.125	1.21889
Dega Edu	0.171	0.000103	0.44	0.125	0.26		0.069444	1.069698
Ebiti	0.171	0.000103	0.6	0.125	0.66	0.19837	0.125	1.87985
Tututi	0.171	0.000103	0.7	0.46	0.4	0.2267	0.069444	2.023473
Elana Kereri	0.125	0.000103	0.7	0.125	0.66	0.15869	0.069444	1.838614
Bentina Mogasa	0.135	0.000165	0.33	0.125	0	0.04192	0.333333	0.965418
Turo Bedenota	0.1525	0.000085	0.44	0.25	0.18	0.19837	0.208333	1.425389
Alge	0.171	0.000085	0.04	0	0.76	0	1	1.971085

**Column heading numbers represent the following*:

(1) Temperature variance, (2) Rainfall variance, (3) Distance from water source, (4) Distance from school, (5) Distance from health centre, (6) Slope in degrees, (7) Women population density, (8) Sum of normalised scores

Climate Change Vulnerability Map

Mapping the climate change and vulnerability of pastoral women of the *woreda* is the final objective of the study as per the previously summarised and calculated index. The given values between 0 and 1 indicated statistically on the index would be converted into a map and displayed with their measurable results – the highest vulnerability index results are for the areas with high vulnerability of climate change and variability. This means that the distance they travel to school, health centres, slopes they cross over, along with climate variation, determine the result represented on the map. A single parameter is not used – instead all distance maps (distance from water sources and school, and poulation density) are included. The *kebeles* in the central parts of the *woreda* have relatively better acces to education, health centres and water sites, while settling on plain elevation, as a result of which they have less vulnerability. Alge and Gola *kebeles* are relatively less vulnerable as depicted on Table 2. The climate change vulnerability map is indicated in Figure 14.

Figure 14. **Women pastoralists' climate change vulnerability map of Fentalle Woreda**

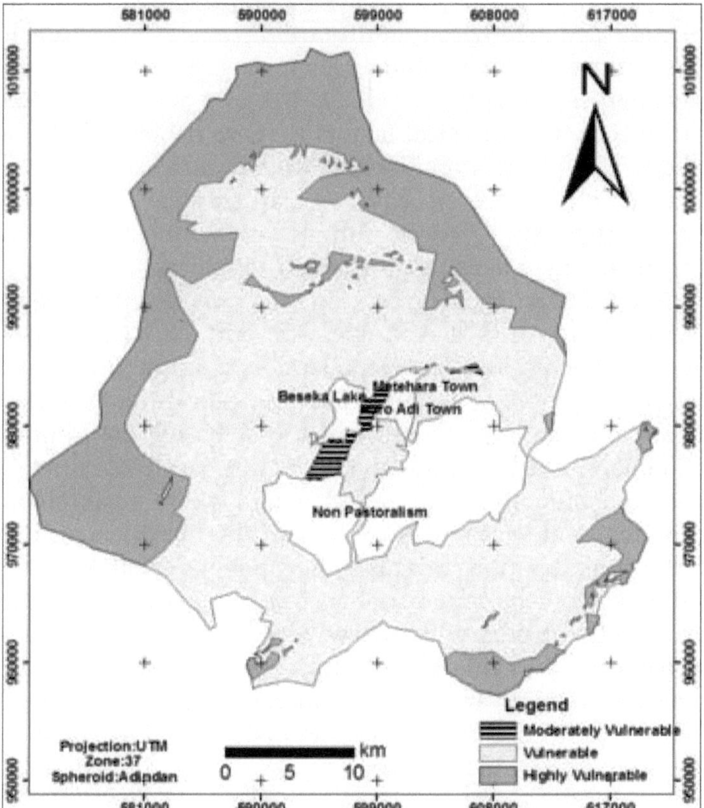

Source: Author's research analysis results

Based on the analysis and multiple indicators, the *kebeles* in the northern and western parts of the *woreda* are more vulnerable to climate change and variability than others. Pastoralists in these *kebeles* suffer from climate variability more than other pastoralists in other *kebeles* with limited availability of infrastructure facilities that makes the coping strategy unsuccessful. Haro kersa, Kobo, Ebiti and Tutut (see Table 2) are among the *kebeles* where the effect of climate changes is higher. Distances from water sites weigh more for all *kebeles* as the climatic elements are insignificant and have slight difference from *kebele* to *kebele*. Hence, pastoral women in remote areas of the *woreda* suffer more as water sites are found at long distances from their current locations._

Conclusion

This paper has used GIS-based multi-criteria analysis approach to assess the vulnerability of climate change and variability on pastoralist women in Fentalle *Woreda*. It has also used the MCA framework to incorporate physical factors that are believed to have impact on the lives of the pastoral women in the *woreda*.

Pastoral women's access to basic social services and their perceptions, feelings and understandings about their physical environment have also been incorporated. Pastoralist women's vulnerability index for each *kebele* in the *woreda* has been constructed. This, in turn, has helped to put on a map and differentiate the highly vulnerable or moderately vulnerable *kebeles*.

Previously, the degree of vulnerability of the pastoral communities due to climate change was coarsely mapped based on limited climatic elements, excluding other spatial factors rarely conducted in Ethiopia and specifically for Fentalle *Woreda*, where pastoralism is a dominant economic system. A review of the literature on the subject under consideration reveals that previous studies have incorporated all socio-economic variables but lack mapping, which is a potential tool for vulnerability assessment.

The approach used in this paper helps researchers and policy makers to easily identify the most vulnerable areas and gives emphasis for spatial factors when, for instance, they plan a community-based development project.

The paper also helps the concerned groups to spot priority areas and pastoralist community groups like pastoral women during climatic hardship.

The findings of the study can be categorised into two groups:

The first group is related to the situation of the climatic factors in the study area. Considering, for instance, the rainfall records, one can conclude that no climate change or sign of climate change has been observed in the *woreda*, given the uniform rainfall intensity data at hand for the last 39 years. This holds true if and only if the recorded value is considered. However, scarcity of water is a fact of life which has been turning the lives of the pastoralist women in the *woreda* from bad to worse. This is simply because there is a great temporal variation. It is the regular distribution of the rainfall as well as the duration – beginning and ending of the rainy season – that makes the rainfall in the area normal. Sometimes, the rain starts early and stops early, or it falls in higher intensity but for limited months. That is what has made the recorded rainfall intensity stay without decreasing for a long time. The other possible factor will be the value of the temperature. The temperature increases from decade to decade. In this case, it makes the pastureland emit more evaporation, thus making the environment dry, although the rain may not have decreased in intensity.

The second analysis group of vulnerability relies on basic social services when expressed spatially. Distance from water, schools and health centres has been considered, assuming other factors constant. The vulnerability index obtained from all these values indicates that the water source is the main challenge for pastoralists under extreme climate as they travel a long distance in search of water. Under harsh climatic conditions the market value of the livestock falls by half (50 per cent) and at this time pastoralist women suffer more than men as the women are less likely to subsidise their income by other means like working as a daily labourers. Crossing a steep slope to reach schools and health centres is also another big challenge facing women under extreme climatic conditions. [In] post-climate change *Bona* conditions are expected to have more adverse impact on their animals than the changed *Genna* conditions.

Limited access to water, pasture and basic social services makes pastoral women of the *woreda* more vulnerable to climate-induced changes. Under extreme climate conditions women-headed households suffer most as they cannot move frequently to other places in search of water and pasturelands as men-headed households do.

Climate variability and change has a proportional relationship with these factors. The peripheral areas are highly vulnerable due to previously indicated factors; that is, distances from water sources, schools, health centres and others. The central pastoralist *kebeles*, excluding the urban residents, are moderately vulnerable except for some locations.

There is a need to develop an adaptation and mitigation strategy to help the pastoral communities adjust themselves to the changing environment. Based on the findings of the study the following recommendations can be made:

- There is a need to involve the pastoralist communities in the formulation and implementation of environment policies and strategies.
- Pastoralist communities should be provided with water at the time of environmental hardship.
- Mass education and increasing people's access to information on climate change and vulnerability should be provided.
- Women's livelihood strategies to reduce the economic burden that can be created by climate change and variability ought to be diversified.
- Basic social services ought to be made accessible to the people.

Annex 1

Table 3: **Sample size taken from each kebele**

Kebele	**Total Female Population**	**Per cent**	**Sample Size**
Haro kersa	828	0.02	17
Ebiti	792	0.02	16
Dega edu	672	0.02	13
Kobo	1120	0.02	22
Bentina Mogasa	1016	0.02	20
Gelcha	1053	0.02	21
Elalana kereri	977	0.02	20
Tututi	2005	0.02	40
gola	823	0.02	16
Alge	850	0.02	17
Ledi fake	1091	0.02	22
Kenifa	785	0.02	16
Sarana weba	817	0.02	16
Gederana kube	640	0.02	13
Gara dima	716	0.02	14
Dire seden	1126	0.02	23
Turo bedenota	877	0.02	18
Gidara	1094	0.02	22
Total	17282	0.02	346

References

Aguilar, L. 2009. Training manual for gender and climate change. San José, Costa Rica: Masterlitho S.A. Accessed from: www.gender-climate.org. (accessed date, 26 December, 2011)

Abebe Tadege. 2007. Climate change national adaptation programme of action (NAPA) of Ethiopia. Addis Ababa: National Meteorological Service Agency of Ethiopia.

Airine, L. D; Homocianu, D. 2009.*The geographical dimension of DSS applications*. Tomul, LVI ŞtiinŃe Economice .

Ayalew G. 2009. When pastoral commons are privatised: Resource deprivation and changes in land tenure systems among the Karrayu in the Upper Awash Valley Region of Ethiopia. Proceedings of the 16th International Conference of Ethiopian Studies , edited by Svein Ege, Harald Aspen, Birhanu Teferra and Shiferaw Bekele. Volume 2, Trondheim.

Building Nigeria's Response to Climate Change (BNRCC). N.D. Geender and climate change.toolkit. Draft version 1.0. Accessed from:www ccaps.robertstrausscenter.org/system/research_items/pdfs/51/original. (accessed date on 11July 2011)

Cavallo, A and Norese, M.F 2000. GIS and multicriteria analysis to evaluate and map erosion and landslide hazards..52th Meeting of the European Working Group. Assessed from http://www.mii.lt/MCDA-52/paper_an.htm. (accessed date 13 April 2011)

Central Statistical Agency. 2007. Population and housing census. Summary and statistical report of 2007 population and housing census. Accessed from http://www.csa.gov.et/pdf/Cen2007_firstdraft.pdf (accessed date 05 January 2011)

Getnet, F. 2011. GIS and remote sensing sased assessment of climate change impacts on food security, the case of Central Rift Valley and adjacent highlands. *Ethiopian Journal of Development Research,* vol.33. Addis Ababa.

Indian Counsel of Social Science Research (ICSSR). 1983. *Survey of research in physical geography*. New Delhi, Concept Publishing.

International Center for Tropical Agriculture. 2011. Climate change, agriculture and food security (CCAFS). Proposal for CGIAR Research Program 7 Accessed from www.*ccafs.cgiar.org*/.../2011/agro biodiversity-address-climate-change-agri (accessed date on 19 December 2011)

International Crop Research Institute for the Semi Arid Tropics (ICRISAT) 2009. Quantitative assessment of vulnerability to climate change. Accessed from: www.Icrisat.Org/What-We-Do/Impi/Training-Cc/October-2-3-2009/Vulnerability-Analysis-Manual.Pdf. (accessed date 15 May 2010)

International Fund for Agricultural Development (IFAD). 2009. Livestock thematic papers: Tools for project design, livestock and climate change. Accessed from: www.Ifad.Org/Lrkm/Index.Htm. (accessed date on 07 September 2011)

Inter Governmental Panel on Climate Change (IPCC).1995. Second assessment climate change report. Accessed from: http://www.ipcc.ch/pdf/climate-changes-1995/ipcc-2nd-assessment/2nd-assessment-en.pdf. (accessed date 07May 2010)

---.2001. IPCCC third assessment report on climate change 2001: Impacts adaptation and vulnerability. Published for the Intergovernmental Panel on Climate Change by UNEP/WMO. Cambridge: Cambridge University press.

---.2007. Climate change. Accessed from: www.Ipcc.Ch/Publications_and_Data/Publications_And_Data.Htm. (accessed date on 07 May 2010).

Jotoafrika. 2011. Adapting to climate change in Africa. Accessed from: http://www.eldis.org/vfile/upload/1/document/1104/JotoAfrika_Issue%206.pdf jato Africa. (accessed date on 23 March 2011).

Kinyangi, J. et al. 2009. Vulnerability to climate change and climate variability in the Greater Horn of Africa: International Livestock Research Institute. Accessed from: www.ilri.org/user/228. (accessed date 04 July 2011)

Lapucci, A, Lombardo,S. and Petri, M. 2005. .A KDD based multi criteria decision making model for fire risk evaluation. Pisa, Italy. Accessed from: www. itcnt05.itc.nl/agile_old/Conference/.../23_Alessandra%20 Lapucci.pd. (accessed date 13 October 2011)

Least Developed Countries Expert Group (LEG). 2002. Annotated guidelines for the preparation of national adaptation programmes of action. Accessed from: unfccc.int/files/cooperation_and_support/ldc/application/.../annguide.pdf. (accessed date 20 August 2011)

Malczewski, J. 1999. GIS and multicriteria decision analysis. New York: John Wiley & Sons.

Mokarram, M. and Aminzadeh, F. 2010. GIS-based multicriteria land suitability evaluation: Using ordered weight averaging with fuzzy quantifier: A case study in Shavur Plain, Iran. The international Archives of the Photogrammetry, *Remote Sensing and Spatial Information Sciences*, Vol. 38. Accessed from: www.Isprs.Org/Proceedings/Xxxviii/Part2/Papers/9_Paper.Pdf. (accessed date on 16 October 2011)

Ministry of Agriculture Ferderal Democratic Republic of Ethiopia. 2011. Forests, rangelands and climate change in the near east region.General outline of country report. Accessed from: www.Fao.Org/Forestry/29286-0bb36f1f012f34032146f810ca33684bf.Pdf. (accessed date on 13 September, 2011)

Muluneh, W. and Demeke, H. 2011. Climate change and variability. Its impact on rural livelihood, local coping and adaptation strategies in Wore Illu Woreda North Eastern Ethiopia. *Ethiopian Journal of Development Research*. Volume 33, No.1. Addis Ababa.

Nori, M. and Davies, J. 2007. Change of wind or wind of Change? Climate change, adaptation and pastoralism. Prepared for the world initiative for sustainable pastoralism. Nairobi: IUCN. Accessed from: http://cmsdata.iucn.org/downloads/c__documents_and_settings_hps_local_settings_application_data_mozilla_firefox_profile.pdf. (accessed date on 10 March 2011)

Olson, J. , Rubin, D., and Wangui, E. 2010. Gender, agriculture and climate change: A regional analysis for USAID/East Africa. Accessed from: http://www.usaid.gov/our_work/crosscutting_programs/wid/pubs/Gender_Agriculture_Climate_Change_East_Africa.pdf. (accessed date 13 April 2011)

Shimelis Beyene and Dafa Gudina. 2009. Reviving a traditional pasture management system in Fentale, East Central Ethiopia . *Journal of Ecological Anthropology*.Vol.13, no. 1. Accessed from: http://shell.cas.usf.edu/jea/PDFs/VOl13beyene.pdf. (accessed date 20 August 2011)

Stanton, E., Cegan, J., Bueno, R. and Ackerman, F. 2011. Estimating regions' relative vulnerability to climate damages in the CRED Model. Stockholm Environment Institute. From http://www.sei-international.org/mediamanager/documents/Publications/Climate/SEI-WorkingPaperUS-1103-v2.pdf. (accessed date on 25 December, 2011)

Temesgen, D. Hassan, R. and Claudia. R. 2008. Ethiopian farmers' vulnerability to climate change across regional states IFPRI Discussion Paper 00806. Accessed from: Http://www.Ifpri.Org/Publication/Measuring-Ethiopian-Farmers-Vulnerability-Climate-Change-Across-Regional-States. (accessed date 16 August 2011)

United Nations Children's Fund (UNICEF). 2008. Our Climate, our children, and our responsibility: The implications of climate change for the World's Children. UK climate change report. Accessed from: www.unicef.at/fileadmin/medien/pdf/Climate_Children_Responsibility.pdf. (accessed date on 4 August 2011)

United Nations Development Programme (UNDP). 2008. Gender and climate change: Impact and adaptation. Asia-Pacific gender community of practice annual learning workshop. Regional gender team Negombo, Sri Lanka. Accessed from: www.snap-undp.org/elibrary/Publications/GenderAndClimateChange.pdf. (accessed date on 11 August, 2011)

——.2009. Resource guide on gender and climate change. Accessed from: www.uneca.org. (accessed date 11 August 2011)

United Nations Framework Convention on Climate Change (UNFCCC). 2006. Background paper on impacts, vulnerability and adaptation to climate change in Africa for the African workshop on adaptation implementation of decision 1/CP.10 of the UNFCCC Convention. Accra.

World Initiative for Sustainable Pastoralism. 2010. Building climate change resilience for African livestock in Sub-Saharan Africa. Eastern and Southern Africa Regional Office, Nairobi. Available from: www cmsdata.iucn.org/downloads/resilence2.pdf: (accessed date 15 December 2011)

Chapter 6

Women Pastoralists and Climate Change Impacts in Kilosa District, Tanzania

*Venosa A. Mushi**

Abstract

This report is a result of a study undertaken to investigate how climate change affects women pastoralists in Kilosa District in Tanzania. The study adopted a cross-sectional research design, in which case-structured questionnaires were administered to a sample of 50 respondents. Four key informants were also interviewed. Quantitative data were analysed using a Statistical Package for Social Sciences computer software. The qualitative data were analysed through factual and logical interpretation of the findings.

Results show that although both men and women pastoralists are negatively affected by climate change and variability, women are more vulnerable. This is mostly due to inequalities between men and women in terms of household gender division of labour, ownership of resources and decision-making. The major effect of climate change as revealed in this study was poverty, as reported by 76 per cent of the respondents. It is recommended that the government should emphasise awareness creation on climate change adaptation practices as well as gender balance, especially in the household division of labour, ownership of resources and decision-making.

*vmakauki@yahoo.co.uk

Introduction

Background to the Study

Global climate change, according to the World Bank (2007b), is expected to have disproportionate negative impact on developing countries, in particular on agricultural productivity in Africa, south Asia and parts of Latin America. It has been predicted that climate change will result in more frequent and severe extreme weather events, such as droughts and floods, with increasing risk to natural resources dependent livelihoods (Watson, Zinyowere, and Moss 1998; Heather, Ford, and Ford 2010; Shemsanga, Omambia, and Gu 2010). The climate change impacts have brought about changes in traditional workload for women and generally affected their social-economic position (Kirkbride and Grahn 2008). The specific climate change aspects which are considered in this study are droughts and floods.

According to the United Nations Convention to Combat Desertification –UNCCD (2007), stress and hardship rise for everyone in times of scarcity of natural resources, but women are most burdened with the increased workload as they struggle to compensate for the resource scarcity. For instance, it is reported by the UN (2008) that, in times of water scarcity, women shoulder a bulk of responsibility and they are more than twice as likely as men to collect water. However, women nowadays struggle for economic opportunities to sustain their livelihoods in times of hardships. For instance, it has been discovered in Ethiopia that during droughts, pastoral communities, particularly women, indicate some shifts in economic activities by dealing in petty trade and cross-border trade (GebreMichael and Kifle 2009).

Women, in the course of their daily tasks, have developed an intimate knowledge of the prevailing climatic conditions. However, their knowledge and capabilities have not been fully recognised, and they often face social and economic exclusion in development issues (Kipuri and Ridgewell 2008). This paper investigates how climate change affects women pastoralists in Kilosa District in Tanzania and thereby informs those policy makers who are interested in formulating relevant climate change adaptation policies.

Problem Statement

Livestock in Tanzania provides a potential contribution to the economy of the country, as the livestock sub-sector generates over a quarter of agricultural GDP (URT 2012). Despite this contribution to the national economy, pastoralists, including women pastoralists, have not been given due consideration and support by the government. For instance, the government recently decided to evict pastoralists who had invaded Ihefu Valley within the Usangu Basin in Mbeya District without giving them any assistance (Paavola 2003; Tenga *et al.* 2008).

Although various studies have been conducted on climate change (Paavaola 2003; Salick and Byg 2007; Tarhule 2007; Nasreen 2008), information with strong focus on women pastoralists in sub-Saharan Africa and particularly in Tanzania is scarce. Thus, the objective of this paper is to explore the impact of climate change on women pastoralists in Kilosa District in Tanzania. Specifically, the paper aims to:

- firstly, identify the historical occurrence of droughts and floods events and their impacts on the livelihoods of pastoralists' community in the study area;
- secondly, identify climate change and variability coping and adaptation strategies in the study area; and
- thirdly, to determine how men and women are involved in household division of labour, decision-making and resource ownership.

Literature Review

Effects of Climate Change on Women Pastoral Livelihoods

Climate change and its consequences receive much attention in the public debate. It is thought that weather extremes such as drought, floods and storms will occur more frequently in the future. Rising temperatures will favour agents of tropical diseases, or will speed up their development (Orindi and Murray 2005; Stern 2006), and probably also contribute to their spread into new areas (Henson 2006). Countries in the tropics that depend primarily on rain-fed farming and/or on pastoralism are thought to be particularly vulnerable to the effects of climate change. For example, the Intergovernmental Panel on Climate Change (IPCC 2007) predicted that, in some of these countries, agricultural production could decline by as much as 50 per cent by 2020. Likewise, the UNEP (2006) has estimated that by 2025, about 480 million people in Africa could be living in water-stressed areas. Climate change is the biggest long term threat against the possibilities to end poverty especially in aspects of hunger and malnutrition. This makes adaptation to climate change a fundamental driver of development in agriculture, including livestock, food security and other efforts to achieve the MDGs (IPCC 2001; Howden *et al.* 2007; Tarhule 2007).

Climate change affects different groups of the society, including pastoralists, differently. The livelihoods of pastoralist communities are highly dependent on natural resources, hence are affected by the impacts of climate change. However, the degree of vulnerability to climate change between and within pastoral communities varies according to factors such as age, sex, wealth rank, economic engagement, ownership of different livestock species and geographical location. This implies heterogeneity in adaptation and difficulty of replication or scaling up of successful adaptations across the community. On the other hand, it is encouraging that pastoralists have accumulated and continue to accumulate deep-rooted knowledge and experience in adapting to the ever-changing environment caused by climate change, and they are flexible (GebreMichael and Kifle 2009).

It is apparent that women have different positions and face different challenges relative to men in the coping and adapting to climate change. According to Tipilda and Kristjanson (2008) women and men livestock keepers typically face different livelihood opportunities and constraints in managing livestock as well as in coping with various challenges. The challenges include poor access to markets, services and technical information, periodic drought and disease, competing resource uses, policies that favour larger-scale producers or external markets and weak institutions. In most systems, women provide labour for the various tasks related to livestock but may or may not control the process of decision-making, particularly over the disposal of animals and animal products. Likewise, women may be involved in production, but may or may not own the means of production, including livestock, land and water. Moreover, Women's Environment and Development Organisation – WEDO (2007, cited in Dankelman *et al.* 2008), states that women, as the majority of the world's poor, are the most vulnerable to the effects of climate change. Thus, during natural disasters, often more women die than men because they are either not warned, cannot swim, or cannot leave the house alone (Dankelman *et al.* 2008). When poor women lose their livelihoods, they slip deeper into poverty due to the increase in inequality and marginalisation they suffer from because of their gender. Consequently, climate change presents a very specific threat to women's security (*Ibid.*).

In the past, pastoralist households in Tanzania, despite living in poverty, owned large stocks of cattle, goats and sheep ranging from 20 to 200 herds per household (Mashindano *et al.* 2011). Currently the country has more than 21.3 million heads of cattle and it ranks third on the list of African countries with high numbers of cattle (URT 2011). However, many of the households have lost their stock due to climate change impacts such as droughts and floods. The consequences of climate change in the country have even taken a gender dimension in which women are seen to be more vulnerable on account of deep-rooted socio-economic and historical barriers (Shemsanga, Omambia and Gu 2010). Climate change and the resulting loss of livestock, water and vegetation has changed traditional workload of women pastoralists. Currently the collection of firewood and water can take longer time than it has been before due to increased drought. Likewise, the preparation of food can be more tedious since ingredients are getting scarce. Also many women, where livestock is lost, have to replace the care for animals by other domestic or incomegenerating activities (Hartmann and Sugulle 2009).

Climate change, along with other social and environmental factors, has altered pastoral livelihoods in various directions, affecting also women's livelihoods, their coping strategies and their social position. Increasing poverty is among the effects of climate change in pastoral societies like the Maasai.

Among the poorest groups of pastoral families, further increase of poverty due to the decline of the natural resource base induced by climate change can lead to disruption of families. This is so because men migrate to towns to look for jobs, while the women are left behind to take care of the children and the remaining livestock. Usually this situation is desperate for the women and children, because of not having more than a few goats left behind. However, as vulnerability to climate change depends on control of financial, physical, natural, human and social capital, and because women typically have less access to and control over these resources than men, they are likely to have lower adaptive capacities.

Illiteracy hinders women's access to information to improve their livelihood enterprises and speed of recovery from a climatic disaster or constrain their options for livelihood strategies. Illiteracy limits an individual's ability to take up opportunities such as employment and inhibits access to information or technical advice that could help them adapt to climate change. Consequently, women's economic insecurity increases more than men's; they take longer to recover from economic loss than men; gender barriers create an obstacle to involvement in reconstruction work and natural disasters. These often lead to a sustained increase in the workload of women, for example, due to male out-migration, reduced access to resources; more time spent travelling to collect water and search for fuel wood (Lambrou and Piana 2005).

Pastoralists' Traditional Adaptations to Natural Disasters

Natural disasters, like floods and droughts, are not new phenomena. For centuries, pastoralists and farmers have adjusted their livelihoods to periods of resource scarcity (McPeak 2006). However, the combination of unsustainable resources use and intensifying climate change threaten to overwhelm the existing adaptive capabilities. Pastoralists traditionally reduce risk from variations in resource availability through various adaptations. These include livestock movement and migrations to track forage and water availability, herd diversification to multiple species (although spreads risk across species), and social programmes such as stock associations and wealth re-distribution from wealthy to poor (*Ibid.*).

Pastoralism has existed in East Africa since 3000 BC (Spear and Waller 1993). While the early Maasai of Tanzania and southern Kenya did cultivate at times in addition to herding livestock, a series of droughts in the eighteenth and nineteenth centuries reinforced specialised Maasai livestock herding under the difficult conditions (Homewood and Rodgers 1991). Historically, complex pastoral exchange networks insured against catastrophic losses by scattering herds amongst far-flung associates (Homewood and Rodgers 1991).

Traditionally, pastoralists typically employ multiple herd management techniques to buffer their losses in difficult times. For instance, Maasai pastoralists of Tanzania herd a mixture of cattle, goats and sheep. These complementary herd species allow livestock to take maximum advantage of available resources in different ecological niches. Mixed herds also ensure that the herd owner is buffered against species-specific disease outbreaks (Galaty and Johnson 1990; Reid, Galvin and Kruska 2008). The ratio of species herded depends on cultural preferences, environmental parameters like availability of grass and water and the personal choices of the herders themselves (Cooke 2007). However, it is important to note that in years following drought, the proportion of livestock held in the small stock (sheep and goat) herd will increase because small stock reproduce at 2 - 4 times the rate of cattle, one – to – four kids per year as opposed to one birth every year and a half - to two years for cattle (de Leeuw *et al.* 1991). So smallstock are particularly useful and important after droughts and other disasters. Another importance of keeping small stock like goat is that goats are frequently sold for cash, given as gifts, or slaughtered for food or ceremony because the amount of money and food generated by a single goat is optimal for day-to-day transactions (*Ibid.*).

A study conducted by African Climate Change Fellowship Programme –ACCFP (2010) on gender, pastoralism and climate change, vulnerability and adaptation in northern Kenya, found that pastoralists are diversifying their livelihoods to cope with climate change and climate variability. Alternative coping strategies include remittances from relatives, transformation of pastoralists into agro-pastoralists, especially in areas situated next to water sources, sale of animals and animal products, seeking employment (mostly casual labour), engaging in small businesses, for example, selling firewood, charcoal and selling local alcohol brew (especially more women are involved in the latter). A lot of women are involved in commercial weaving and increasingly in running small shops while men are involved in carpentry, masonry and migration to towns for casual labour. Mobility is a common form of coping with drought or floods. However, emerging conflicts over use of resources is restricting the freedom of mobility, thereby enhancing vulnerability.

Similarly, the UNCCD (2007) reports how pastoralist women in Kenya cope with climate change variability through economic diversification. These women have identified alternative ways of earning some extra income, such as engaging in small-scale bead-work enterprises and growing dryland crops like drought-resistant maize and local cereals, mostly for home consumption. They also earn income by making necklaces, bangles and earrings from beads bought from the nearby shopping centre and selling them at the local market.

Another mechanism used by the pastoral communities during resource scarcity is moving from one place to another. For instance, as a consequence of resource scarcity in pastoralist-dominated regions like in Shinyanga, Arusha and Dodoma, a huge influx of pastoralists with large herds of livestock to Morogoro region has been observed (Paavola 2003). In addition to movement of animals, pastoralists in drought-prone areas have adapted to reserve pasture for weak stocks such as sick, young and lactating animals. Such a method would enable them to survive during the drought season and reduce deaths of weak individuals (Shayo 2006). Moreover, the repeated occurrence of climate-related disasters has forced some pastoralist societies in Tanzania to reduce the numbers of their herds as a coping mechanism. Interestingly, some pastoralist societies have even strategised into paid employments in cities and cultivation of food crops (*Ibid.*).

Nonetheless, these ways of coping with climate change impacts seem not to be sustainable especially with the increase in frequency and severity of extreme events like drought and floods. Also involvement in so many activities as a source of livelihood signifies stress in the pastoral system.

Theoretical Framework

In explaining the relationship between women pastoralists and their environment and the way [they] are impacted by climate change, this paper uses the Gender/women Environment and Sustainable Development theory. According to this theory, women are more compassionate and have nurturing capacities in relation to environment than men; so much so that women are seen as privileged managers of the environment (Braidotti *et al.* 2004). As women (especially pastoralists) daily deal with the environment and nature, they become the victims of climate change impacts. In relation to this women are trapped in a certain disproportionate gender-based division of labour (*Ibid.*).

There are three schools of thought describing the theory of gender/women environment and sustainable development, namely: Descriptive Liberal Approach, Eco-Feminism, and Feminist Environmentalist. This study derives its theoretical framework mainly from the first one – the Descriptive Liberal Approach. The Descriptive Liberal Approach describes the way in which women interact with their physical environment and natural resources, such as land, water and forests (Dankelman 2003). Pastoralist women have direct interaction with the environment as they are providers of water, firewood and food to the household. In this case, women depend solely on their immediate environment and its available resources to ensure survival of their household members and livestock. As they directly extract resources like firewood, building poles, pasture, water and herbal medicine, they have an accumulated knowledge of managing the resources and the ways of coping during instances of scarcity (*Ibid.*). It has been illustrated in the UNCCD (2007) that as keepers of traditional knowledge, women pastoralists make an important contribution towards the sustainable management of land and natural resources the world over.

Climate change impacts differently on women and men depending on their different social and economic roles in the household and community level (Baten and Khan 2010). Deterioration of the environmental resources due to climate change renders women pastoralists victims since they are the most unprivileged people in the pastoral society. The victimisation results from cultural norms that govern household division of labour, decision and ownership of resources. Furthermore, women pastoralists are entrusted more with domestic roles without involvement in community roles. This does not give women the authority to control resources and to decide their livelihoods (Baten and Khan 2010). Consequently these hinder them from coping and adapting effectively to climate change.

Methodology

The study was conducted in Kilosa, which is among the six districts of Morogoro Region in Tanzania. The district was selected based on the fact that livestock keeping is an important economic activity in the area following the large influx of pastoralists from Ihefu wetland and other northern parts of Tanzania (Paavola 2003; Tenga *et al.* 2008). Another reason for selecting Kilosa District is that conflict over resource use between sedentary agriculturalists and nomadic pastoralists occurred in the past, resulting in the death of 30 people (Maganga, Odgaard and Sjaastad 2007). The study has adopted a cross-sectional research design as it allows data to be collected at single point without repetition (Babbie 1990). According to Bryman (2004) the design provides a snap-shot of ideas, opinions and information on the issue under study. The relevance of this design is based on its ability to provide in-depth information. The study population included all the pastoral community households in Kilosa District, whereby the pastoralists' households formed the unit of analysis. The respondents were either the husband or the wife in a household.

Two wards (Rudewa and Madoto) and one village from each ward with pastoralists were purposively selected. The villages involved were Twatwatwa and Mbwade in Rudewa and Madoto wards, respectively. The selection of the wards and villages was based on the availability of pastoralist households. The total number of pastoralists in Kilosa District is estimated to be 3,600 households. A sample of fifty (50) respondents (25 male and 25 female) were selected from the households by using stratified simple random sampling. Although this sample size is small, triangulation was used to enrich the information obtained from the respondents by interviewing key informants, conducting Focus Group Discussions (FGDs) and the researcher's observations. According to Kamuzora (2006), triangulation helps to substantiate information obtained from different sources. Hence, it limits personal and methodological biases and enhances study reliability. In this study, four (4) key informants, including the District Veterinary Officer (DVO), the District Livestock Field Officer (DLFO), the Village Executive Officer (VEO) and the Village Chairperson, were purposively selected. These were elites who could provide required information on women pastoralists and climate change impacts. It was also observed during the interview that the pastoral population was homogeneous; hence even a small sample size was enough to come up with sound conclusions.

The study has used both primary and secondary sources of data. The instruments used for primary data collection were structured questionnaire for respondents, interview guides for key informants and a checklist of questions for the FGDs. Three FGDs each composed of eight members were conducted. Two FGDs comprised men only while the remaining one consisted of both men and women. This was because only few women were ready to participate in the FGDs due to cultural barriers which limit women's participation in public discussions.

Documentary review of relevant literature from journal articles, books and other public documents with information on pastoralist women and men, livelihood issues and climate change was also done. Quantitative data were entered using the Statistical Package for Social Sciences (SPSS) computer programme, and analysed using descriptive statistics such as cross-tabulation with chi square test, frequencies and percentages. Qualitative data were content analysed, where the use of factual and logical interpretation of the study findings was employed.

The paper starts with an introductory section, which includes the background to the study and statement of the problem sub-sections. A review of literature, along with theoretical framework of the study, is covered in the next section which is followed by discussion of the methodology. The following section deals with the discussion of the findings, and the last section provides the conclusion and policy implications of the study.

Findings and Discussion

Demographic Characteristics of the Study Population

Discussion of the characteristics of a study population is of prime importance since it helps in data interpretation. The respondents were asked about their age, sex and years spent in school, as these variables were deemed important in interpretation of the data. The variables are as discussed in the following subsections.

As there was an equal number of male and female respondents involved in the study, it helped to get information which is not gender-biased. Men and women were given equal chances to present their views on the issue under study. However, there was an exception in FGDs where very few women were ready to participate in public discussion as mentioned in the methodology part.

Table 1: **Distribution of respondents, by background variables (n = 50)**

Variable	Frequency	Per cent
Sex		
Male	25	50
Female	25	50
Age group		
<20	5	10
21-30	8	16
31-40	22	44
41+	15	30
Level of education		
Never attended school	30	60
Primary	18	36
Secondary	2	4

Source: Author's field survey, 2011

Furthermore, the age distribution of the respondents was as shown on Table 1 where it ranges from below 20 to above 41 years old. The majority (44 per cent) of the respondents were between the age of 31 and 40 years. This age group is usually the middle age which is energetic. The age group from 41 years and above constituted 30 per cent of the respondents and this group is expected to have wider knowledge on the issue of climate change and variability. Demographers and other social scientists have special interest in the age structure of a population because several social relationships within the community depend on age (Mbonile 1993). The age structure of a population determines several demographic events like education cycles and the like.

In relation to education the study findings show that 60 per cent of the respondents had never attended formal education, while 36 per cent and 4 per cent had attended primary and secondary education, respectively. This shows the actual picture of pastoral communities, especially the Maasai people who hardly value formal education. The results also depict the actual situation in most rural areas of Tanzania, where the majority had never attended secondary education. Similar cases were reported in Kenya where 84 per cent and 91 per cent of pastoralists interviewed in Turkana and Mandera, respectively, were illiterate (DFID 2010). Another case has been reported in Australia that in indigenous communities, as much as education may be valued, it is not the central focus of their existence nor their social and cultural way of life. In such cultures, the members of the community see little use in investing their resources and time in education (Chhetri and Baker 2005). It has been argued that education, among other things, brings about gender equality, improved socio-economic status and poverty reduction at large, which then contributes to privileged performance of women across many of the Millennium Development Goals (Fentiman, Hall, and Bundy 1999).

Education is an important asset in enabling poor people to come out of poverty especially when the situation is expected to be exacerbated by ensuing climate change. It is associated with greater access to information on climate change, improved technologies and higher productivity. Education is also assumed to increase climate change awareness, hence promoting adaptation. Avila and Gasperini (2005) argue that illiteracy strongly correlates with poverty and hunger and is particularly high among rural women. They add that it hinders rural development and limits opportunities to improve livelihoods and promote gender equity. The big number of the respondents who had never attended school suggested that the majority of the pastoralists in the study area are indeed denied some opportunities to improve their livelihoods.

Occurrence and Effects of Droughts and Floods

Before engaging in the discussion on the occurrences of droughts and floods in the study area it is worth to know the understanding of the study population on climate change. Study findings show that almost all (98 per cent) of the respondents were knowledgeable on the term climate change. More than two-thirds (70 per cent) of the respondents explained climate change as the increase of occurrence of droughts. This meaning was given by the majority because there had been recurrent drought in the area as revealed during focus group discussions. On the other hand, 20 per cent of the respondents explained climate change as an increase in temperature and lack of rainfall, while 6 per cent of the respondents explained it as occurrences of both floods and droughts. The remaining 4 per cent said that they had no idea on the meaning of climate change.

Although the respondents seemed not to understand the literal meaning of climate change, most of them have an idea on the indicators of climate change. In line with this, Kelbessa (2007) argues that indigenous people's interpretation of climate change depends on individual observations, experiences and local cultural framework. Salick and Byg (2007) add that local interpretation of climate change is often much more diverse and specific to a particular local region. Cross-tabulation shows that all the respondents who could not give the meaning of climate change were men. This shows that men are not much concerned with climate change impacts as compared to their female counterparts.

The results of the present study reveal that there have been frequent droughts and floods in Kilosa District as reported by 94 per cent of the respondents. These findings are supported by the key informants interviewed during the study who also gave the various periods of droughts and floods. According to the key informants, floods occurred in 1977/78, 1998 and 2009. It was discovered during focus group discussions that the most destructive floods were those of 2009. The FGD participants claimed that the destruction from the floods caused massive deaths of animals and a big loss of pasture. It was, however, surprising to note that the 1998 floods, which were also known as *El Niño*, were considered a 'blessing' to pastoralists. The reason they gave for this is that there was increased pasture and water for their livestock brought by the *El Niño* rains. Despite this 'blessing', as termed by the pastoralists, *El Niño*, as the FGD participants argue, has affected the health of some animals, like goats, which suffered from foot rot diseases, leading to the death of some.

Apart from the effects of floods, the pastoralists were also negatively affected by droughts in Kilosa. The study findings show that since the last decade, the frequency of droughts has been increasing. Droughts in Kilosa District have also been reported in other areas apart from the studied wards (Rudewa and Madoto) by Mung'ong'o and Mwamfupe (2003). The respondents in the study area claimed that climate change, which in this study is considered in terms of droughts and floods, affected them in different ways. The magnitude of the effects is presented on Table 2. As shown on the table, there were two major effects as identified by the respondents and verified by the key informants. These were: poverty, which was mentioned by more than three-quarters (76 per cent) of the respondents, and death of livestock, as mentioned by 70 per cent of the respondents. The next two effects, namely, scarcity of pasture and water resources, and conflicts between pastoralists and crop growers were mentioned by 56 per cent and 54 per cent of the respondents, respectively. As reported by Muung'ong'o and Mwamfupe (*Ibid.*), conflicts over resources, particularly land and water between sedentary agriculturalists and nomadic pastoralists, have been on the increase. Conflicts over resources in Kilosa caused a serious effect to the people. For instance, in December 2000, thirty people were killed at Rudewa-Mbuyuni Village during clashes between crop cultivators and livestock keepers (Maganga, Odgaard and Sjaastad 2007).

The other effects which were mentioned by the minority of the respondents include pupils' school drop-out and outbreak of diseases which were mentioned by 8 per cent and 2 per cent of the respondents, respectively.

Table 2: **Distribution of respondents, by effects of climate change***

Effect	Frequency**	Per cent
Poverty	38	76
Death of livestock	35	70
Scarcity of pasture and water resources	28	56
Conflicts between pastoralists and crop growers	27	54
Pupils' school drop-outs	4	8
Outbreak of diseases	1	2

Source: Author's field survey, 2011

* Climate change in this study is explained in terms of droughts and floods.

** The numbers do not add up to 50 due to multiple responses.

Similar results have been reported by Hazell, Oram and Chaherli (2001) who contend that droughts bring significant losses of livestock and push many herders into poverty. The authors add that poverty holds back investments in better natural resources management. Inadequate investment in natural resources management leads to scarcity of water and grazing land as well. Furthermore, it has been reported by the United Nations' panel of climate experts that Africa is highly vulnerable to the impacts of climate change because of factors such as widespread poverty, recurrent droughts, inequitable land distribution and over-dependence on rain-fed agriculture (IPCC 2001). The dimensions of poverty mentioned by the respondents in the study area are both income and non-income dimensions of deprivations. As revealed during focus group discussion, the pastoralists claim to experience declining income due to the effect of climate change. On the issue of non-income dimension of deprivation, as explained by Moore and White (2003), it involves a lack of access to basic needs and social services. During focus group discussions it was revealed that non-income poverty manifests itself in the lack of food. Furthermore, the outbreak of diseases without reliable treatment which then intensifies death of livestock also falls under dimension of non-income deprivation of poverty.

The climate change effects mentioned are inter-connected, forming a viscous cycle of poverty. Taking an example of death of livestock, this acts as a cause of poverty, but on the other hand, it could also be caused by poverty. When pastoralists are deprived of income, they cannot, for example, buy medicine for vaccinating or treating their livestock. The problem of declined income affects women more adversely because they do not own household income as revealed in the study area. Women are also more affected since they are the ones who make sure that the family eats, but have no command over the available resources.

Coping and Adaptation Strategies towards Climate Change and Variability

This part specifically concentrates on floods and droughts which are the main aspects of climate change considered in this study. The respondents mentioned various short and long term measures in dealing with floods and droughts according to their experiences. The responses, as presented in tables 3 and 4, show only the affirmative responses while the rejecting or negative ones are not indicated. The numbers and consequent percentages appear to be small showing that the majority of the respondents were not practising any coping or adaptation strategy.

The findings (Table 3) reveal that there were several strategies used by the pastoral households in coping with climate change impacts in responding to the shortages of pasture and water due to droughts. The findings show that 30 per cent of the respondents mentioned shifting of animals to new areas as a coping strategy; 22 per cent of the respondents mentioned charcoal production; 14 per cent mentioned selling animals and selling firewood; while 10 per cent mentioned providing supplementary feed and selling of local brew; and 8 per cent mentioned urban migration. Among the mentioned coping strategies, two coping strategies – which are shifting of animals to new areas and selling of charcoal – were significant at P = 0.031and P = 0.088, respectively. This implies that these two coping strategies are the most practised by the pastoralists in the study area. This results into environmental problems such as pollution and land degradation. The environmental pollution also promotes climate change due to greenhouse gases.

Table 3: **Respondents' coping strategies towards droughts and floods (n = 50)**

Strategy	**Male**		**Female**		**Total**		**X^2 Value**
	No.	%	No.	%	No.	%	
Shifting animals to new areas	11	22	4	8	15	30	.031
Charcoal production	3	6	8	16	11	22	.088
Selling of animals	4	8	3	6	7	14	.684
Selling of local brew	4	8	1	2	5	10	.157
Providing supplementary food	4	8	1	2	5	10	.157
Urban migration	3	6	1	2	4	8	.163
Selling firewood	2	4	1	2	3	6	1.0

Source: Author's field survey, 2011

* The numbers do not add to 50 due to the effect of multiple responses.

Furthermore, the findings show that shifting of animals to new areas was an important coping strategy. However, more men than women practise this strategy. According to focus group discussion participants drawn from Twatwatwa Village, the reason given was that women are not in favour of this strategy because it is the responsibility of women to carry all household utensil and children during shifting while men just move with their animals. Another reason given was that sometimes when men shift with their animals women remain at home taking care of the family and young stock. In addition to that, women also find themselves assuming some traditionally men's responsibilities when their husbands are away from their homes. All these increase the burden on women.

In line with the above findings it is clear that due to climate change, the pastoralists' way of life is undergoing great transformation and the trend is moving towards abandoning pastoral life. Pastoral production system depends on the availability of natural resources which are sensitive to climate change. What complicates their situation is that pastoralists do not have a diversified pool of resources to draw from, which makes them more susceptible to outside pressures such as climate change. The situation of pastoralists as well as future uncertainties of pastoral production has made many households opt for some sort of diversification in order to cope with the climate change impacts.

The kind of diversification can be agro-pastoral, together with the incorporation of trees into farming systems (agro-forestry). The agro-forestry kind of farming systems has been identified as a promising approach that has already had far-ranging impacts in many agro-pastoral areas in Africa (World Bank 2007b). This system, apart from providing food to the pastoralists, it is also expected to provide fodder for the livestock.

Charcoal burning from wood has been a new strategy to cope with economic losses from climate change and other external factors like banning of livestock in wildlife conservation and water catchments areas in Ngorongoro in Arusha and Ihefu wetland in Mbeya regions, respectively. Charcoal production became a fallback option for poor pastoralists who are unable to carry on with their traditional way of life. Charcoal-making, as a non-livestock income-generating activity, is widely practised in the study area. According to focus group discussion participants, charcoal-making is the most important option for generating alternative income. This is because the activity does not require much initial capital and there is also free access to tree resources to produce charcoal. Hence, charcoal-making is the second important coping strategy towards climate change impacts in the study area.

As evident from Table 4, pastoralists in the study area use various adaptation strategies towards climate change impacts. The strategies mentioned by the respondents include crop-farming (30 per cent), selling of animals (22 per cent), rural-urban migration (20 per cent), fencing reserve pasture, and shifting to new areas (18 per cent). Pastoralism is becoming more complex than ever before due to the loss of livestock as a consequence of climate change. Those herders who lose their livestock tend to opt for another way of living by becoming agro-pastoralists where they produce maize as their main food crop and beans and vegetables as minor crops. They also rear smaller numbers of livestock and usually keep the milk from animals for household consumption and the surplus for selling. However, with the increasing frequency of drought events, there is no evidence that agriculture can be a viable adaptation to climate change risk reduction strategy or production alternative to pastoralism unless there is good government intervention, appropriate technology and serious implementation of policies.

Table 4: **Respondents' adaptation strategies towards droughts and floods (n = 50)**

Strategy	**Male**		**Female**		**Total**		**X^2**
	No.*	%	No.	%	No.	%	**Value**
Crop farming	7	14	6	12	13	26	0.747
Selling animals	5	10	6	12	11	22	0.733
Migration	7	14	3	6	10	20	0.508
Fencing reserve pasture	4	8	5	10	9	18	0.684
Shifting to new areas	6	12	3	6	9	18	0.269

Source: Author's field survey, 2011

* The numbers do not add to 50 due to the effect of multiple responses.

Selling animals is one of the adaptation strategies which seem to be practised by the majority of the pastoralists in the study area since it was mentioned by 22 per cent of the respondents. In the FGD held in Mbwade Village, it was reported that the households sell old, and sick animals which have given birth two and more times. The income obtained from selling of animals is used for household immediate consumptions, investment in crop farming and some use the income as a start-up capital when they migrate to towns. In this way, family members who succeed in getting employment in the urban areas support their families in the rural areas. Nonetheless, the trend leads to turning these rural families dependant on the support of extended families members in town, while the pastoral mode of living deteriorates. Selling livestock as an adaptation mechanism can leave households more vulnerable to poverty in the long-run.

A study conducted in Somalia found that many herders who have lost all their livestock decide to start alternative economic activities to deal with dwindling resources and asset basis (Abdi 2007), among which farming, charcoal production or trading are the dominant ones. Others move to the urban centres where they try to make their daily living. This study has noted that the dominant coping strategies are shifting animals to new areas and engaging in charcoal production. Although these strategies help reduce the impacts of climate change to the pastoralists, it is important to underscore the negative impact on the environment and the permanent loss of livelihoods which could result from charcoal production and shifting of animals to new areas.

Gender Participation in Terms of Division of Labour, Decision-Making and Resource Ownership

One of the objectives of this paper was to determine gender participation in terms of division of labour, decision-making and resource allocation, and climate change and variability coping and adaptation strategies. The results from the focus group discussions and key informant interviews show that women participate in milking, selling of milk and other animal products, fetching firewood and water, food preparation, cleaning of the surroundings, caring for the children, the sick and old people while men participate in grazing and selling of animals (Appendix 1). These findings imply that the roles of men and women are different based on the household division of labour. Study results reveal that women in the study area shoulder a lot of household activities compared to men. This is in line with what is reported by the UN (2008) that women shoulder the bulk of responsibility for collecting water when none is available on the premises.

Decision-making in the Maasai household is also based on gender. The findings show that the decisions on various aspects are done by men. These aspects include decision on children's education, buying food for daily and future use, size of the herd to be maintained, selling and buying animals, migration, use of land and involvement in income-generating activities. Although decision to engage in income-generating activities is done by men, both men and women are involved in various income-generating activities to cope with the impacts of climate change.

In one of the FGDs, a 47 years old man reported concerning the ownership of resources in the household as follows:

> In our society a man owns almost everything within the household, including livestock, children, and land and can buy things like radio, bicycle, mobile phone and motorcycle because he also has control over the household income while a woman owns only water-keeping vessels and other kitchen ware. When a woman is married she comes to the house empty-handed, hence nothing belongs to her in a household. However, nowadays a Maasai woman can own a mobile phone in rare cases. This is possible for a woman, especially if she sells milk. Likewise, a woman should not be given a room to make decision in the household. She should be submissive to the husband.

It was discovered in the study area that there was no joint decision-making between men and women in a household. Decisions made by women were negligible. This could be a constraint to women pastoralists in coping with climate change impacts when it comes to a point of deciding whether to apply a certain coping strategy like migrating to urban areas for employment in order to earn their living as an alternative to pastoralism. Women cannot decide to migrate because they, by default, carry the responsibility of taking care of the family; thus during drought they become more vulnerable and easy targets as they have no broad alternatives. As a result, when men cope by migrating to towns, women remain behind with a big burden of taking care of the whole family alone. In the FGDs held at Mbwade and Twatwatwa villages, it was reported that women engage in selling firewood, trading milk and milk products, like milk butter and ghee, as well as vending traditional herbal medicine and local brew.

Another aspect unilaterally decided by men is sale of animals and animal products like animal skin and milk. Selling of these items is solely decided by men, but women are responsible for purchasing and preparing food. In cases of drought, the women have no power to decide to sell the animals in exchange for food; again, even when the money is there, they have no choice or the power to decide on how to use the available money. Likewise, while the purchase of household items, like a radio set, largely rests with the men, the decisions on purchase of household necessities, like kitchen items, rest mainly with the women. These findings support the argument made by URT and UNICEF (1990) and Mung'ong'o, Kwasila and Madulu (1990) in the study conducted in Tanzania that women are not provided equitable voices in decision-making since they are accustomed to subordination to men. It has also been asserted by GebreMichael and Kifle (2009), that the poor with fewer livestock and less voice in decision-making in the community are more vulnerable, adding that women household heads are usually classified as poor.

The findings on ownership of resources within the pastoralist society, the Maasai in this case, show that men own the more valuable household resources and assets like cattle, radio, bicycle, motorcycle, household income, and [even the] children. The only major resource owned by women alone was kitchen ware, which is obvious because they are the ones who take care of cooking in the household. Items like mobile telephones and radio sets could be owned by both men and women, depending on who bought the item, particularly so in the case of mobile telephones. Despite women's important roles as producers and household managers, as they are usually involved in all household chores, they are often marginalised when it comes to allocation and ownership of resources (Appendix 1). They lack direct access to resources and this consequently reduces their capacity to cope with and adapt to climate change impacts. It has been suggested by Mushi and Makauki (Forthcoming) that the government needs to have programmes for awareness-raising on resource allocation and distribution to benefit both men and women. This, however, should not remain at national level, but to go down to the community or grassroots level. There is a need for the community to be aware of the importance of mutual ownership of resources equally between men and women. When a resource belongs to a person, it is more likely that person handles it with care due to the feeling of ownership. Needless to say, family resources are very helpful in coping with climate change hazards such as droughts and floods. Families with enough resources at their disposal can easily survive the consequences of droughts and floods.

Avila and Gasperini (2005) note that improving women's access to land, agricultural inputs, financial services and skills, among others, works in favour of empowerment, equality and better lives for women and their families. The authors further point out that gender inequality increases the risk of the female malnutrition, disease incidence, child mortality and other family problems that erode human capital and further reduce women's access to assets. Due to unequal access to resources at the household level, women lacking of ownership of productive resources like household income, cattle, and use of land renders them incapable of coping with their changed roles. For instance, women cannot engage in income-generating activities such as running small businesses due to the lack of financial capital as it was seen that the household income was owned by men. Despite this fact, as Mashindano *et al.* (2011) report, women shoulder the burden of undertaking income-generating activities single-handedly. This makes women more exposed to the impacts of climate change.

It has further been contended by the authors that women's ability to meet various living costs through farming has fallen as fertile land is in short supply. This is due to the impacts of climate change – mainly frequent floods and prolonged droughts. The situation, however, is exacerbated more by the lack of ownership of land which, culturally but not legally, cuts across among women all over Tanzania. It has further been argued that meanwhile, as women have taken on more and more of men's share of the responsibility for family maintenance, the cash costs they face in sustaining their families have risen sharply since 2000 (*Ibid.*).

In the United Nations (2003) Millennium Development Goals, Goal 3 aims at improving gender equality and empowering women. The high priority accorded to the goal represents a global affirmation of women's rights and gender equality as core values of development. In fact, attaining MDG-3 is put a pre-requisite for the attainment of the other MDGs as elaborated in the Global Monitoring Report of 2007. The report highlights that attempting to achieve the MDGs without promoting gender equality will both raise the costs and decrease the likelihood of achieving the other goals (World Bank 2007a; World Bank 2007b). The reverse is equally true, that achievement of MDG-3 depends on progress made on each of the other goals. Improving gender equality also influences poverty reduction and growth directly through women's greater labour force participation, productivity and earnings. Thus, females generally need to be empowered so that their contribution in development can be recognised. Women empowerment can be possible by providing them with appropriate education as well as giving them the opportunity to participate fully in various development issues (Mushi and Makauki 2010).

It has been reported by the UNDP (2003) that in the national reports for the MDGs, gender perspectives and women's concerns are reflected only in some countries' MDGs but not in others. For instance, women concerns are not visible in Goals 7 and 8 which are on environment and development cooperation, respectively. The exclusion of women in the discussion of the vast majority of national reports on MDGs suggests that women are still being seen in terms of their vulnerabilities, and cast in their traditional roles as mothers or victims rather than as actors in development. What is more, gender mainstreaming in the reports still looks patchy.

Conclusion and Policy Implication

This paper has assessed the impact of climate change and variability on pastoral women in Kilosa District in Tanzania. It has tried to address the question of how women pastoralists are affected by climate change and variability in the study area. The study has been guided by three specific objectives:

The first objective has been to identify the historical occurrence of drought and flood events and their impacts on the livelihoods of pastoralists' community in the study area.

Secondly, the study has aimed at identifying climate change and variability coping and adaptation strategies in the study area.

Thirdly, the study has sought to determine gender participation in terms of division of labour, decision-making and resource ownership.

Education level of the study population seemed to be very low, suggesting challenges in adopting new technologies and adaptation and coping strategies. The main effect of droughts and floods in the study area is poverty, which also exacerbates the other effects. Moving animals to new areas and engaging in crop farming are the two main coping and adaptation strategies towards climate change impacts. Inequality has been found in the gender-based division of labour, decision-making and ownership of resources, where women are disadvantaged in all areas.

It is recommended that there should be awareness creation programmes on the importance of education to both men and women. This will help increase literacy levels of the community, hence making it easier for them to get employment and adapt to modern technologies. Since the major effect of climate change in the study area is poverty, there is a need for government intervention to rectify the situation.

Firstly, the intervention can be in the form of educating people on climate change mitigation measures. For instance, instead of depending solely on livestock keeping, pastoralists can engage in agro-pastoralism. Due to gender inequalities in household division of labour, decision-making and resource ownership, there is a need first to mainstream gender into any community programmes.

Secondly, special programmes focusing on gender issues are necessary for the whole community. Such initiatives could be providing training programmes in gender equality and its importance in improving people's livelihoods.

It is also important to involve the pastoralists in interventions measures concerning their livelihood so that they have a sense of ownership and hence, give the necessary cooperation for the successful accomplishment of the intervention. Involvement of pastoralists in such interventions is expected to reduce resistance to change. When they are involved it will be much easier for them to agree, for instance, to reduce the size of herds as one of the mechanisms to lessen land degradation.

The government should as well strive to assist the pastoralists in improving the marketing of their animal and animal products. This requires training them in entrepreneurial skills. It is expected that doing so will increase the income of the pastoral communities and hence assist in achieving poverty reduction.

Appendix 1

Household gender division of labour, resource ownership and decision-making

Variable	**Response by gender**	
Division of labour	Female	Male
Milking	√	
Selling milk	√	
Selling of other animal products e.g. skin	√	
Fetching firewood	√	
Fetching water	√	
Food preparation	√	
Cleaning of the surroundings	√	
Caring of children, the sick and the old people	√	
Grazing of animals		√
Selling of animals		√
Purchasing of food	√	√
Resource ownership		
Household income		√
Cattle		√
Kitchen ware	√	
Motorcycle		√
Radio		√
Mobile telephone	√	√
Livestock products (milk, ghee, skin, hide)	√	√
Decision-making on:		
Children's education		√
Buying food for daily use and future use		√
Size of the herd to be kept		√
Selling and buying animals		√
Migration		√
Engagement in income-generating activities		√

References

Abdi, M. 2007. No more 'grass grown by the spear': Addressing land-based conflicts in Somaliland: Land-based conflict project Somaliland report. Hargeisa: Academy for Peace and Development, p. 60.

African Climate Change Fellowship Programme (ACCFP). 2010. Gender, pastoralism and climate change: Vulnerability and adaptation in Northern Kenya. Final technical report. (October), p.9,

Avila, M. and Gasperini, L. 2005. *The MDGs and sustainable rural development in sub-Saharan Africa: Challenges and implications for education for rural people*. Rome: FAO.

Babbie, E. 1990. *Survey research methods*. Belmont California: Wadsworth Publishing Company, p.395.

Baten, M. A. and Khan, N. A. 2010. *Gender issues in climate change discourses: Theory versus reality*. Dhanmondi, Dhaka: Unnayan Onneshan-The Innovators,

Braidotti, R.; Charkiewicz, E.; Hausler, S. and Wieringa, S. 2004. *Women, the environment and sustainable development: Towards a theoretical synthesis*. New Delhi,: Vedams eBooks. p. 220.

Bryman, A. 2004. *Social research methods* (2nd ed.). Oxford: Oxford University Press.

Chhetri, N. and Baker, D. P. 2005. The environment for literacy among nations: Concepts, past research and preliminary analysis. Background paper prepared for the Education for All Global Monitoring Report 2006: Literacy for Life. Paris: UNESCO.

Cooke, A. E. 2007. Subdividing the Savanna: The ecology of change in Northern Tanzania ecology. PhD dissertation. Chapel Hill: University of North Carolina.

Dankelman, I. 2003. Gender, environment and sustainable development: Theoretical trends, emerging issues and challenges review paper. Santo Damingo: United Nations International Research and Training Institute for the Advancement of Women (INSTRAW), P. 60.

Dankelman, I.; Alam, K.; Ahmed, W.A.; Gueye, Y. D.; Fatema, N. and Mensah-Kutin, R. 2008. *Gender, climate change and human security: Lessons from Bangladesh, Ghana and Senegal*. WEDO, P. 73.

de Leeuw, P. N.; Semenye, P. P.; Peacock, C. P. and Grandin, B. E. 1991. *Productivity of cattle and small stock Maasai herding*. Addis Ababa: International Livestock Centre for Africa, pp. 83-101.

Department for International Development (DFID) UK. 2010. Gender, pastoralism and climate change: Vulnerability and adaptation in Northern Kenya. Final technical report. Available from: http://start.org/download/accfp/omolo-final.pdf: (accessed on 12th November, 2011).

Fentiman, A.; Hall, A. and Bundy, D. 1999. School enrolment patterns in rural Ghana: A comparative study of the impact of location, gender, age and health on children's access to basic schooling. *Comparative Education*, 35 (3): 331-349.

Galaty, J. and Johnson, D. 1990. Pastoral systems in global perspective. In *The world of pastoralism: Herding systems in comparative perspective*, edited by J. G. Galaty and D. L. Johnson, New York: The Guilford Press, pp.1-32.

GebreMichael Y. and Kifle M. 2009. *Local innovation in climate-change adaptation by Ethiopian pastoralists*. Addis Ababa: PROLINNOVA-Ethiopia and Pastoralist Forum Ethiopia, p. 25.

Hartmann, I. and Sugulle, A. J. 2009. Candlelight for health, education & environment: Report on the impact of climate change on pastoral societies of Somaliland. Djibouti: Candlelight for Health, Education & Environment, p. 62.

Hazell, P., Oram, P. and Chaherli, N. 2001. Managing livestock in drought-prone areas of the Middle East and North Africa: Policy Issues. In *Food and agriculture in the Middle East: Research in Middle East economics,* edited by Hans Löfgren, vol. 5. New York: Elsevier Science.

Heather, E. T.; Ford, L. B. and Ford, J. D. 2010. Climate change and food security in Sub-Saharan Africa: A systematic literature review. *Sustainability*, 2, 2719-2733.

Henson, R. 2006. *The rough guide to climate change: the symptoms, the science and the solutions*. London: Rough Guides Ltd.

Homewood, K. and Rodgers, W. 1991. *Maasailand ecology: Pastoral development and wildlife conservation in Ngorongoro, Tanzania*. Cambridge: Cambridge University Press.

Howden, S. M.; Soussan, J. O.; Tubiellos, F. N.; Chhetri, N.; Dunlop, M. and Meinke, H. 2007. Adapting agriculture to climate change. PNAS, vol. 104, no. 50 (December 11), pp.19691–19696 Available from: www.pnas.org_cgi_doi_10.1073_pnas.0701890104. (accessed on 10th February, 2011).

Intergovernmental Panel on Climate Change (IPCC). 2001. Climate change impacts, adaptation and vulnerability: Contribution of Working Group II to the third assessment report of IPCC. Cambridge: Cambridge University Press.

——.2007. Working Group contributions to the fourth assessment report. Fourth assessment report. Climate change 2007. Synthesis report. Summary for policy makers, p. 22.

Kamuzora, F. 2006. A synthesis of utilization of information band communication technologies in Tanzania tourism system. PhD dissertation. Bradford: University of Bradford.

Kelbessa, W. 2007. Climate change impacts and indigenous coping strategies in Africa. Paper prepared for the International conference on Riding on a Moral Storm. The Global Challenge of Climate Change: Science-Economics-Ethics-Politics. Greifswald University, p. 34.

Kipuri, N. and Ridgewell, A. 2008. *A double blind: the exclusion of pastoralist women in the East and Horn of Africa*. London: Minority rights Group International, p. 40.

Kirkbride, M. and Grahn, R. 2008. Survivor for the fittest: Pastoralism and climate change in East Africa. *Oxfam Briefing Paper*, No. 116. Oxford: Oxfam International, p. 47.

Lambrou, Y. and Piana, G. 2005. *Gender: The missing component in the response to climate change*. Rome: FAO and UN.

Maganga, F.; Odgaard, R. and Sjaastad, E. 2007. Contested identities and resource conflicts in Morogoro Region, Tanzania. Who is indigenous? In *Conflicts over land and water in Africa*, edited by Derman B. Odgaard R. and Sjaastad, E. Oxford: James Currey Ltd., pp. 202-214.

Mashindano, O.; Kayunze, K.; da Corta, L. and Maro, F. 2011. Agricultural growth and poverty reduction in Tanzania 2000-2010: Where has agriculture worked for the poor and what can we learn from this? Working paper no. 208.

Mbonile, M. J. 1993. Migration and urban development in Tanzania. In *Population migration and changing world order*, edited by Gould W.T.S and Findley A.M. Oxford: John Willey & sons Ltd.

McPeak, J. 2006. Confronting the risk of asset loss: What role do livestock transfers in northern Kenya play? *Journal of Development Economics*, 81 (2): 415-437.

Moore, M. and White, H. 2003. Meeting the challenges of poverty and inequality. In *States, markets, and just growth: Development in the twenty-first century*, edited by Atul Kohl, Chung-in Moon and Georg Sorensen. Tokyo: United Nations University Press.

Mung'ong'o, C. and Mwamfupe, D. 2003. *Poverty and changing livelihoods migrant Maasai pastoralists in Morogoro and Kilosa Districts, Tanzania*. Dar es Salaam: Mkuki na Nyota Publishers, p. 41.

Mung'ong'o, C. G., Kwasila H. L. and Madulu N. F. 1990. A *Socio-economic assessment of North Eastern Parts of Tanzania*. Dar es Salaam: University of Dar es Salaam.

Mushi, V. A. and Edward, M. C. Forthcoming. Challenges and success of gender budgeting initiatives: A case of Tanzania. *Accountancy and Business Review Journal*.

Mushi, V.A. and Makauki, A. F. 2010. Contribution of socio-cultural and economic factors to girls' schooling in Sub-Saharan Africa. *African Affairs* 27, 205-234.

Nasreen, M. 2008. Impact of climate change on food security in Bangladesh: Gender and disaster perspectives. Paper presented at an International Symposium on Climate Change and Food Security in South Asia, held in Dhaka, 25th – 30th August, 2008.

Orindi, V. A. and Murray, L. A. 2005. *Adapting to climate change in East Africa*: A strategic approach. London: International Institute for Environment and Development, p. 25.

Paavola, J. 2003. Livelihoods, vulnerability and adaptation to climate change in the Morogoro Region, Tanzania. Centre for Social and Economic Research on the Global Environment. Working Paper EDM. Norwich: University of East Anglia.

Reid, R. S., Galvin, K. A. and Kruska, R. S. 2008. Global significance of extensive grazing lands and pastoral societies: In *Fragmentation in semi-arid and arid landscapes*, edited by Galvin K. A. et al. Dordrecht: Springer, pp.1-24.

Salick, J. and Byg, A. 2007. Indigenous people and climate change. Available from: http://tyndall.webapp.uea.ac.uk/publications/Indigenouspeoples.pdf. (accessed on 23rd May, 2011).

Shayo, C. M., 2006. Adaptation, planning and implementation: Agriculture and food security. UNFCCC African Regional Workshop on Adaptation, 21st – 23rd September, 2006, Accra, Ghana. Dar es Salaam: Vice President's Office.

Shemsanga, C., Omambia, A. N. and Gu, Y. 2010. The cost of climate change in Tanzania: Impacts and adaptations. *Journal of American Science*, 6 (3): 182-196.

Spear, T. and Waller, R. 1993. *Being Maasai: Ethnicity and identity in East Africa*. London: James Currey Ltd.

Stern, N. 2006. The Stern review on the economics of climate Change. HM Treasury, UK: Cambridge University Press. Available from: http://www.hm-treasury.gov.uk/independent_ reviews/stern_review_economics_climate_change/sternreview_index.cfm. (accessed on 13th January, 2011).

Tarhule, A. A. 2007. Climate information for development: An integrated dissemination model. *Africa Development*, Vol. XXXII, no 4, 127-138, CODESREA.

Tenga, R.; Mattee, A.; Mdoe, N.; Mnenwa, R.; Mvungi, S. and Walsh, M. 2008. A study on options for pastoralists to secure their livelihoods in Tanzania. Current policy, legal and economic issues. Volume one: Main Report, p. 138.

Tipilda, A. and Kristjanson, P. 2008. Women and livestock development: A review of the literature. International Livestock Research Institute (ILRI) Innovation Works Discussion Paper 01-09. Nairobi: ILRI. Available from: http://www.ilri.org/Link/Files/InnovationWorks/WomenLivestock_Background_Literature_Review_Final.doc. (accessed on 12th November, 2011).

United Nations (UN). 2008. *The Millennium Development Goals report: End poverty 2015. Make it happen.* New York: United Nations, p. 56.

United Nations Convention to Combat Desertification (UNCCD). 2007. *Women pastoralists: Preserving traditional knowledge: Facing modern challenges.* Bonn: United Nations.

United Nations Development Programme (UNDP). 2003. *Millennium Development Goals: National reports, a look through gender lens.* New York: UNDP.

—–.2006. *Marine and coastal ecosystems and human well-being: A synthesis report based on the findings of the millennium ecosystems assessment.* Nairobi: UNEP, p. 64.

United Republic of Tanzania (URT). 2011. Development plan and budget estimate for 2011/2012: Speech of the Ministry of Livestock and Fisheries Development. Available from: http://www.mifugo.go.tz/documents_storage/2011-7-29-17-26-53_bajeti_2011_2012.pdf. (accessed on 2nd April, 2012).

—–.2012. Livestock. Available from: http://www.tanzania.go.tz/livestock.htm#top. (accessed on 2nd April, 2012).

URT and UNICEF 1990. *Women and children in Tanzania. A situation analysis.* Dar es Salaam: Government Printer, p. 169.

Watson, R. T., Zinyowere, M. C. and Moss, R. H. 1998. *The regional impact of climate change: An assessment of vulnerability.* Cambridge: Cambridge University Press, p. 517.

World Bank. 2007a. *Global monitoring report 2007. Millennium Development Goals: Confronting the challenges of gender equality and fragile states.* Washington DC: The World Bank.

—–.2007b. *World development report 2008: Agriculture for development.* Washington, DC: The World Bank.

Chapter 7

Gender and Climate Change Adaptation among Agro-Pastoral Communities: Case Study of Chivi District in Southern Zimbabwe

Admire Mare[*]

Abstract

This study discusses the positive and negative survival strategies adopted by agro-pastoral communities in Zimbabwe as they struggle to deal with the vagaries of climate change. It focuses on agro-pastoralists living in Chivi District, who have been on the receiving end of climate change-induced drought and famine for the past couple of years. Looking at these agro-pastoralists as self-aware actors in charge of their lives and destinies rather than defenceless victims, the study discusses the various strategies (both positive and negative) developed by villagers in Chivi District to keep their families afloat, to broaden their livelihood options and to keep their livestock healthy. It discusses the social and economic capital systems they utilise to cope with the harshness of life in Chivi District. It argues that the gendered impacts of climate change are relative and contingent upon factors such as household dynamics and the wider socio-cultural and economic circumstances in which the agro-pastoralists operate. It recommends the implementation of integrated climate change and social policies in order to mitigate the impact of climate change in the area characterised by a fragile ecosystem.

*admiremare@gmail.com

Introduction

> Climate variability and change will result in fundamental alterations to ecosystem structures and functions. These in turn will affect human land-use and livelihoods and have the potential to make pastoralists more vulnerable (Galvin, Thornton, Boone and Sunderland 2004, 1).

There is unanimity among scholars, international think-tanks and non-governmental organisations that:

> ...human activity which takes place within ecological systems, and is not bound to political frontiers and will have negative impacts on the environment and people's well being beyond geographical boundaries if not managed in a sustainable manner (IPCC 2001; WEDO 2007; Ongoro and Ogara 2012;).

Climate change is seen as increasing climate variability characterised by extreme seasonal and annual temporal and spatial variability in rainfall patterns and amounts (IPCC 2007). It results in a variety of direct problems, including increased frequency of extreme weather events, flooding, storms, drought, desertification, increases in sea temperatures, heat and cold waves, the melting of glaciers and permafrost (IPCC 2001; Warren 2006). These developments will have significant ecological, social, economic and political impacts, including effects on food production, water availability, intensifying wildfires, mud-streams, bleaching of corals, changes in epidemic vectors and extinction of pollinators (IPCC 2001; WEDO 2007). Despite the anticipated severe impacts of climate change, the burden will not be homogeneous across countries, gender, race, ethnicity and geographic regions (IPCC 2007; WED0 2008). For instance, men and women will be differentially impacted on and experience different levels of vulnerability to climate change (Denton 2002; WEDO 2007). Similarly, the poorest countries and communities will bear most of the brunt of climate change (IPCC 2007). In the southern African region, including Zimbabwe, the climate has become highly variable and unpredictable (Gandure 2005). The region is prone to extreme weather conditions and is generally expected to get drier and continue to experience more extreme weather conditions, particularly droughts and floods (Mubaya 2011a).

Although we still have much to learn about the earth's climate, given its changes and variability in its natural cycles and the impacts, it has become increasingly evident that subsistence agricultural producers, more so small farmers, stand to suffer the greatest (IPCC 2001; Mubaya 2011a). In particular, scientists believe that as climate change continues to manifest itself across the globe, a 'pastoral crisis' will leave millions of people facing starvation as their livelihoods become precarious (Meier, Bond and Bond 2007). Pastoral communities are those that have a disproportionate subsistence emphasis on herding domesticated livestock (Birch and Grahn 2007). A pastoral system represents a complex form of natural resource management, which Pratt, Le Gall and De Haan (1997) define as the direct interaction between the ecosystem, the social system and the geo-political system. According to Kirbride and Grahn (2008), arid and semi-arid areas (ASALs) constitute some of the harshest and remotest places on earth. Meier Bond and Bond (2007) argue that given this aridity, it is hardly surprising that pastoralists are subject to higher levels of risk than those living in areas where farming is a viable option. In these dryland environments, rainfall can be anywhere from 25mm to 600mm, often varying as much in time and place as in quantity (Bollig and Schulte 1999). It is, therefore, striking that within such environments mobile livestock herders – pastoralists – survive at all (Birch and Grahn 2007; Kirbride and Grahn 2008). Globally, pastoral production systems are thought to support 100 – 200 million mobile pastoralists. Mobility is the core of the pastoral livelihood system, and crucial to managing risk in these harsh and unpredictable environments (Pratt, Le Gall and De Haan 1997; Swift 2003). The movement of livestock according to water and pasture availability allows communities to utilise large areas of rangeland which lack permanent sources of water. Complex systems of natural resource management and negotiated access enable groups to effectively co-utilise and sustain pasture and water resources (Kirbride and Grahn 2008). In the Sahel, for instance, climate changes in temperature and rainfall patterns have reduced the length of the vegetative period and made it difficult to continue the cultivation of traditional varieties of long and short cycle millets (Ongoro and Ogara 2012). Researches conducted in northern Kenya (Omolo 2011; Ongoro and Ogara 2012) and Somalia (Hartmann and Sugulle 2009) highlight the precariousness of pastoral communities.

Problem Statement

Besides case studies from a few countries (Botswana, Mozambique, Namibia and South Africa) commissioned by Heinrich the Boell Foundation in 2010, there is a dearth of studies on the relationship between gender and climate change in southern Africa. The few studies done on gender and climate change have largely been about agricultural communities (WEDO 2007; Mubaya 2011a; 2011b). Still, fewer studies are specifically on adaptation strategies by pastoral communities in Africa (Jost 2002; Hartman and Sugulle 2009; Omolo 2011; Ongoro and Ogara 2012). On the other hand, there are a handful of studies discussing coping strategies deployed by agro-pastoralist women in southern Africa. Cognisant of this lacuna in academic research, the WEDO (2007) reminds us that the positive contributions of women on climate change are not well recognised, studied and documented yet. In fact, women's intimate knowledge about resource management and adaptation skills is not well researched in eastern and southern Africa. Extending this view into the contours of gender and climate change, Dankelman (2002) suggests that there is a need to consider studies of the relationship between gender and the environment, in order to determine the gender dimension of climate change. Research on the link between gender and climate change in countries like Zimbabwe is insignificant and largely speculative. This is despite mounting evidence showing that the country is facing and experiencing the adverse impacts of climate change which are manifesting themselves in increased recurrence of natural disasters in the form of droughts, cyclones and floods (Gandure 2005; Mubaya 2011a & 2011b; Tigere 2011). The situation is even worse in agro-pastoral communities, such as Beitbridge, Chivi, Gwanda, Plumtree and Hwange, where rain-fed agriculture has become difficult (Gandure 2005). Thus far, research conducted in Chivi District has focused on alternative livelihoods strategies by agricultural communities in the face of climate change (Dhliwayo 2007); little scholarly attention has been focused on the impact of climate change on agro-pastoralists. In light of the dearth of studies on agro-pastoralist communities in southern Africa, this research project has sought to examine the impact of climate change and variability on agro-pastoralists in Chivi District, and examined the resultant coping mechanisms and explored the gendered impact of climate change in the area.

Study Objectives

This study has sought to investigate the gendered dimensions of climate change and variability among agro-pastoral women in Chivi District. More specifically, it has assessed and examined: the perceptions of agro-pastoral women on the occurrence and impact of climate change and variability; gender differentiated impacts of climate change and variability; and also documented agro-pastoralist women's historical, traditional and new coping and adaptation mechanisms to climate change and variability.

Importance of the Study

Given the scarcity of studies on gender and climate change in southern Africa, especially focusing on agro-pastoral communities, this study is timely and valuable for policy makers to design appropriate adaptation policies. It also compensates for the gap through contributing to the ongoing debate about gender and climate change. It attempts to show how an understanding of the gendered nature of climate change can help policy makers to avoid implementing gender-blind climate change adaptation policies in Zimbabwe. By focusing on the nexus between gender and climate change within an agro-pastoral context, the study shows how women are both vulnerable and self-aware actors with adaptive capacity within a constraining structural environment. Overall, the study makes an invaluable contribution in the sense that it advocates a nuanced understanding of vulnerability and adaptation capabilities within a rural context. It shows that female-headed, child-headed and grandmother-headed households are particularly vulnerable and, therefore, calls upon government authorities to implement progressive social policies, such as the Zunde RaMambo (Chief's granary), national grain loan scheme, social grants and Basic Education Assistance Module (BEAM) in rural areas. It also highlights that the gendered nature of climate change reminds us that gender mainstreaming rhetoric unaccompanied by real socio-economic empowerment of disenfranchised groups in society engender feminisation of poverty and contributes to negative coping mechanisms, such as prostitution, which in turn lead to feminisation of HIV and AIDS.

Research Methodology

This study was carried out in Chivi District, Masvingo Province in Zimbabwe. Chivi is a semi-arid area which falls in agricultural regions 4 and 5. The district is located in the dry-prone region of the country, north of Mwenezi District and west of Masvingo District. The district's population is estimated to be about one million, consisting of ethnic Karanga people (Dhliwayo 2007). For people in Chivi District, agro-pastoralism is the principle mode of production whereby livestock rearing (cattle, sheep/goats), complemented by communal farming, are the major sources of livelihoods. The district is an agricultural marginal area and has a fragile ecosystem. Chivi District is classified as one of the highly food insecure regions in the country. The annual rainfall for the district ranges from 650 - 800mm with periodic dry spells during the rainy season. The district is characterised by sandy soils, an element which entails low water retention. In the past, the first rains, known locally as the *gukurahundi* or *bumharutsva*, used to come in late September or early October (Dhliwayo 2007). Chivi District has over the years experienced a number of natural disasters such as droughts and cyclone Eline. These natural disasters left a trail of destruction, including gullies, soil erosion and loss of natural bio-diversity. Recurring droughts have also affected the availability of fodder.

Data Collection

Data for this study was collected from four villages (Zivuku, Machanja, Chikanda and Vhudzi) in Chivi District. These were selected using purposive sampling. Purposive sampling is simply selecting groups or categories for the study on the basis of their relevance to the research questions and the theoretical position and most importantly the explanation or account which one is developing. Interviewees for this study were selected using convenience sampling, which is a sampling technique involving the selection of the most accessible subjects (Marshall 1996).

To understand agro-pastoralists' perceptions of climate change and variability, this study has employed a predominantly qualitative research methodology. Qualitative research refers, in the broadest sense, to research that produces descriptive data, people's own written or spoken words and observable behaviour (Taylor and

Bodgan 1984, 5). It allows the researcher to understand perspectives of social actors, to retrieve experiences from the past, and to gain expert insight or information (Babbie and Mouton 2001). The qualitative methods of data collection used include: participatory rural appraisal (PRA) tools (such as observation, historical trend analysis, semi-structured meetings, and unstructured interviews); focus group discussions (FGDs); key informant interviews, and life histories.

As part of PRA, the researcher participated in cattle herding sessions with locals; volunteered in land preparations; and also attended four weekly village meetings. Four focus group discussions (FGDs) consisting of between 6 to 10 people were organised with the assistance of traditional leaders. In coming up with participants of each FGD, factors such as age and gender were taken into consideration. To make communication more effective, the local language (Shona) was used to allow for informed discussions. In order to recruit participants for FGDs and life history interviews, the researcher took advantage of weekly village meetings held at the headman's compound. Through the help of traditional leaders in the area, eight elderly men and women (over 65 years of age) were picked for life history interviews. The group further included agricultural extension officers, staff from prominent non-governmental organisations and teachers.

To supplement data collected from PRA, FGDs, life histories and interviews with key informants, secondary data were obtained by means of desk-top research. This involved a comprehensive review and analysis of existing published and unpublished literature relevant to the study.

Literature Review

Many of the implications of natural disasters on women and men are documented in existing literature (Denton 2001& 2002; Parikh 2007; Petrie 2010). This literature demonstrates that in many societies, vulnerability to natural disasters differs for women and men. Women are often more vulnerable to disasters than men through their socially constructed roles and responsibilities, and because they are poorer (Denton 2002). Following up on gendered research on natural disasters, an emerging strand of studies has begun to document the nexus between gender and climate change in recent years (Denton 2002; WEDO 2007; BRIDGE 2008; Brody, Demetriades and Esplen 2008; Carvajal, Quintero and García 2008). Most of the literature relevant for this particular study focuses on the impact of climate change and gender roles in community adaptation (Ongoro and Ogara 2012; Kirbride and Grahn 2008), traditional strategies used by pastoralists to cope with drought (Mworia and Kinyamario 2008) and climate change-induced conflicts amongst pastoralists in northern Kenya (Omolo 2011). However, two schools of thought are discernible in literature which interrogates the relationship between gender and climate change.

One school of thought argues that women are victims of natural disasters because of the existing gender inequalities prevailing in most patriarchal societies. At the core of this perspective is the view that men and women will be differently impacted and vulnerable to climate change due to the socio-cultural, physiological and socio-economic differences (IPCC 2007). This perspective is based on previous research on poverty and disaster studies where societal cracks and crevices are seen as disadvantaging women. It is believed that the anticipated gender differential impacts would be attributed to existing inequalities, such as unequal access to resources, gendered divisions of labour and decision-making power, which may affect the ability to respond to the effects of climate change (WEDO 2007; BRIDGE 2008). Women, in general, are also responsible for productive tasks such as food collection and energy supply for the household as well as many care-giving tasks, such as caring for the children, the sick, the elderly, the home and assets (Enarson 2008). Furthermore, women are expected to be more vulnerable than men to the impacts of climate change as they are reported to be the majority amongst the poorest and most disadvantaged groups in society (Denton 2002; WEDO 2008).

Furthermore, this school of thought believes that women's relative lack of education and access to information on projected climate change impacts, such as rainfall patterns, or effective natural resource management strategies, like efficient water use, crop diversification or rotational grazing, limit their ability to cope with or adapt to the effects of climate variability and change (WEDO 2008). Studies from gender and disaster research such as the cyclones in Bangladesh and Hurricane Katrina in Louisiana reveal that gender and poverty intersect to make poor women most vulnerable (Nasreen 2000; Anderson 2002; Ariyabandu and Wickramasinghe 2003). Women are, therefore, seen as having a weaker socio-economic status, coupled with limited capacity, to be able to cope with and adapt to climate change (Parikh 2007). It has also been noted that basic needs, such as food, water, health and energy, which are mainly affected by climate change, are the basis of women's livelihoods (Denton 2002). Women in developing countries are responsible for climatically sensitive tasks, such as securing food, water and energy, which ensure food security and household well-being (WEDO 2007). The disadvantaged position of women means greater difficulty in coping with disasters. Overall, climate change is seen as magnifying existing inequalities, reinforcing the disparity between women and men in their vulnerability to and capability to cope with climate change (Mitchell, Tanner and Lussier 2007).

An opposing perspective views women as positive agents of change and contributors to livelihood adaptation strategies (Enarson 2000; O'Brien 2007). This school of thought holds that women are the best managers of natural resources because of their closeness to and high dependence on natural resources; their local knowledge systems are also considered important for climate adaptation (WEDO 2007; Omolo 2011).

Although women may be the most vulnerable, it has also been acknowledged that they function as agents of change in community natural resource management, innovation, farming and care-giving. They, therefore, hold the key to adaptation to climate change (WEDO 2008). For centuries, women have passed on their skills in water and forest management and the management of bio-diversity, among others. It is through these experiences that women have gained valuable knowledge that will allow them to contribute positively to the identification of appropriate adaptation and mitigation techniques given the opportunity (Denton 2002). This perspective argues that women play a major role in natural resource and environmental management, yet their contribution is ignored.

Furthermore, this point of view challenges the dominant narrative which obfuscates the agency of women in disaster management. It sees the view that women are the main victims of climate change as too simplistic. In fact, women are treated as positive agents of change and contributors to livelihood adaptation strategies. Extending this argument further, Enarson (2000) argues that natural disasters could also provide women with a unique opportunity to challenge and change their gendered status in society. For instance, research on recent disasters have shown that women have been willing and able to take an active role in what are traditionally considered 'male' tasks in responding to disasters, as was seen following Hurricane Mitch in Guatemala and Honduras in 1998. In general, women have proved effective in mobilizing the community to respond to disasters, and in disaster preparedness and mitigation (Mitchell, Tanner and Lussier 2007). However, women still own fewer assets than men to recover from natural disasters, and usually do not own land that can be sold to secure income in an emergency. The problems women identify encountering when having to adapt to climate change include: lack of safe land and shelter, lack of other assets and resources, limited access to material and financial resources, lack of relevant skills and knowledge, high prices of agricultural inputs and other materials, and cultural barriers limiting women's access to services (Petrie 2010). Figure 1 sums up the relationship between gender and climate change in developing countries.

Figure 1: **Relationships between gender and climate change**

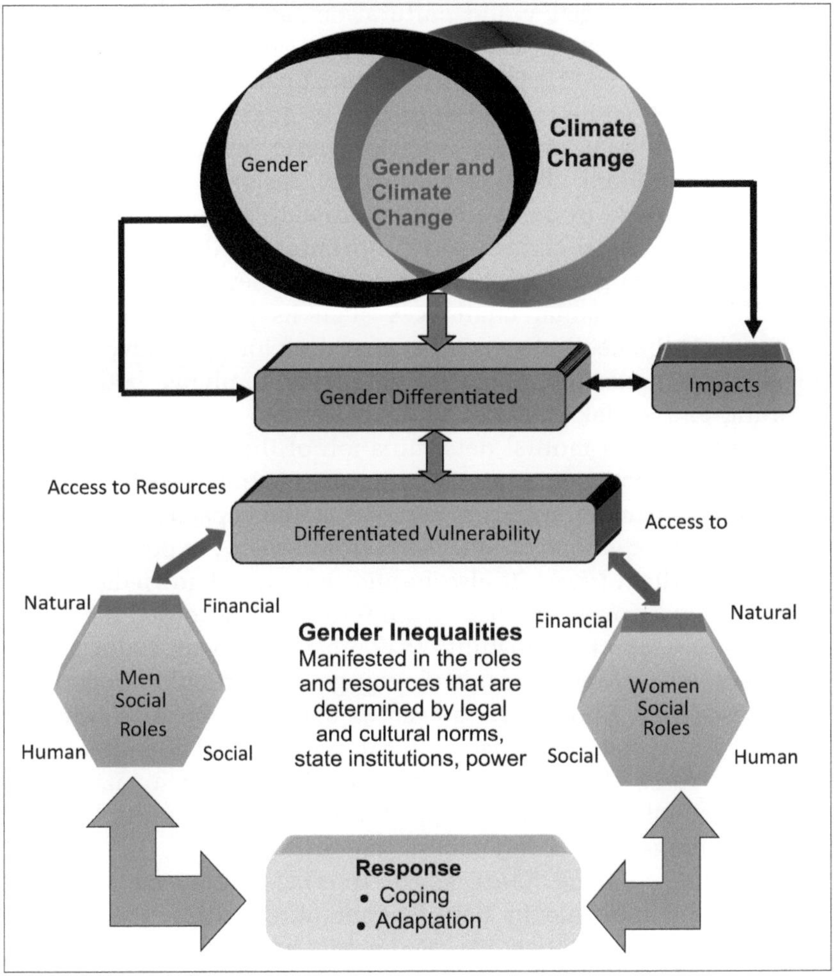

Source: Adapted from Babugura (2010, 30)

Conceptual Framework: Anthony Giddens' Structuration Theory and Norman Long's Actor-Oriented Approach

This study is influenced by Norman Long's (1992) Actor-Oriented Approach and Anthony Giddens' (1986) Structuration Theory which foreground the notion of human agency and life worlds. In particular, the current study seeks to identify and address the ways in which human agency and cultural embeddedness (life worlds) shape the way women pastoralists respond to climate change and variability. Giddens' (1986) theory shares a lot of similarities with the basic tenets of actor-oriented approach as popularised by Norman Long. Both theories proceed from the basic premise that human beings are not blank slates (*tabula rasas).* Instead agro-pastoralists are active subjects in the worlds they inhabit, important players in the politics of social change and contributors to the production of knowledge. Agro-pastoralists use a range of 'devices' to cope with stressful environmental challenges such as climate change and droughts. At the heart of an actor-oriented sociology of development is the characterisation of social action as implying both social meaning and social practice. 'Human agency' and self-organising processes, and the mutual determination of the so-called 'internal' and 'external' factors and relationships are central to social practice (Long 1992). Human agency is defined as the capacity of actors to process their experiences, and learn from others', experiences and to act upon them (*Ibid.*). It also implies a certain knowledgeability, whereby experiences and desires are reflexively accorded meanings and purposes, and the capability to command relevant skills, access resources of various kinds, and engage in particular organising practices (*Ibid.*). Knowledge is seen as emerging out of a complex inter-play of social, cognitive, cultural, institutional and situational elements. It is, therefore, always essentially provisional, partial and contextual in nature, and people work with a multiplicity of understandings, beliefs and commitments (*Ibid,* 18).

According to Long (*Ibid*), social interface situations are more complex and multiple in nature, containing within them many different interests, relationships and modes of rationality and power. The concept of interface helps us to focus on the production and transformation of differences in world-views or cultural paradigms. Interface situations often provide the means by which individuals or groups come to define their own cultural or ideological positions vis-à-vis those espousing or typifying opposing views. On the other hand, Giddens' (1986) Structuration Theory is an attempt to 'put sociology back together again'. Its starting point is that the division between structural and social action approaches is essentially false. This theory's major quest is to reconcile the oft-cited dichotomy between the individual and society. It elucidates the intersections between knowledgeable and capable human agents and the wider social systems and structures in which they are implicated. Giddens (*Ibid.*) argues that human agency and social structure are not two separate concepts or constructs, but are two ways of considering social action. There is a duality of structures so that on one side it is composed of situated actors who undertake social action and interaction, and their knowledgeable activities in various situations. At the same time, it is also the rules, resources, and social relationships that are produced and reproduced in social interaction. Giddens (*Ibid.*) seeks to demonstrate that social structures and social actions have a symbiotic relationship. They are dependent upon each other, but also constrain each other. Therefore, neither social structures nor social actions can exist independently of one another. Giddens (*Ibid.*, 25) defines structuration as "the structuring of social relations across time and space, in virtue of the duality of structure". This means that the structure is implicated in every moment of action - that is at once constraining and enabling - and that, conversely structure was an 'absent' order of differences, 'present' only in the constituting moments of interaction through which it was itself reproduced or transformed.

For Giddens (*Ibid.*, 134), "actors are not inherently predisposed to sustained reasoning or existential reflection on the meaning of their conduct from moment to moment in everyday life". Rather, 'discursive consciousness' emerges at critical times – expected or unexpected. In these circumstances, "actors mobilise their efforts and focus their thoughts on responses to problems which will diminish their anxiety, and ultimately bring about social change" (*Ibid.*, 134-5). Furthermore, social life is a skilled accomplishment on the part of a knowledgeable and capable human subject rather than an automatic response to any 'trans-historical' logic or functional imperative. At the core of the Structuration Theory is the assumption that all human action is performed within the context of a pre-existing social structure which is governed by a set of norms which are distinct from those of other social structures (*Ibid.*). Of paramount importance to this study is the view that the structure and rules are not permanent and external, but sustained and modified by human action.

Against this backdrop, this chapter departs from 'structuralist' approaches to human behaviour and reinvigorates 'human agency' sociological approaches to closely examine how agro-pastoralists deploy *practical consciousness* (knowledge of 'how to go about' the business of social life) and *discursive consciousness* (the ability to tell 'rational stories' about our activities) in order to cope with the impacts of climate change. It specifically seeks to highlight that agro-pastoralists are reflexive and self-monitoring beings, which enables them to respond to changing situations. This conceptual framework offers a key guiding frame in understanding agro-pastoralists as having the ability to invent, cope and improvise within the structures of their routines.

Study Findings and Interpretations

This section presents findings and interpretations of the study under the headings: perceptions on the occurrence and impact of climate change and variability; gender differentiated impacts of climate change and variability and historical, traditional and new coping; and adaptation mechanisms in response to climate change and variability. It then proceeds to summarise, conclude and recommend a number of concrete policy interventions to address the gendered vulnerabilities associated with climate change in Chivi District.

Qualitative data were categorised into three distinct themes as follows:

- Perceptions on the Occurrence and Impact of Climate Change and Variability;
- Gender Differentiated Impacts of Climate Change and Variability; and
- Historical, Traditional and New Coping and Adaptation Mechanisms in Response to Climate Change and Variability.

Perceptions on the Occurrence and Impact of Climate Change and Variability

In the FDGs with participants from Chivi District, climate change was described as one of the major problems they were facing. Both men and women demonstrated awareness of climate variability and change. They concurred that climate change was a recent phenomenon as evidenced by reduced rainfall patterns. Their understanding was mainly based on experience and oral communication. In their own language, they expressed that climate change is associated with abnormal warming of the earth that will lead to changes in the year-to-year seasonal climate.

However, when probed further on the causes of climate change in the area, divergence of opinion emerged. The causes ranged from spiritual (ancestral curses), to earthly sins by human beings, air pollution and deforestation. In the interviews with villagers from Zivuku, Machanja, Vhudzi and Chikanda, participants described climate change as a spiritual and a natural phenomenon. In terms of natural causes, high temperatures and changes in wind movement and moon movements were cited as contributory factors. Most of the interviewees believed that climate change was caused by the wrath of ancestral spirits who have turned their back on Zimbabwe. They accused traditional authorities for failing to take the lead in appeasing ancestral spirits and conducting traditional rites. For instance, most FGD participants pointed out that it has been long since they last held a rain-making ceremony *(mukwerera)* in the area. According to them, *mukwerera* is integral because it entails asking for rain from God and showing gratitude for the rains in the previous season. Participants in Machanja, Vhudzi and Zivuku believed that their failure to hold rain-making ceremonies annually had angered the ancestors and the only way to reverse this tragedy was to appease them.

It is clear that agro-pastoralists try to make sense of what is happening in their environment based on the socio-cultural framework in which they operate (Mubaya 2011a). Most interviewees bemoaned a spike in social ills such as prostitution, high rates of divorce, bestiality, homosexuality and secularisation of society as equally responsible for angering the ancestors. As Long (1992) points out, 'knowledge emerges out of a complex inter-play of social, cognitive, cultural, institutional and situational elements'. Similarly, Mubaya (*Ibid.*) suggests that farmers in Lower Gweru and Lupane also linked the political crisis in Zimbabwe at the time of the research and the decline of social and cultural practices to the variability in climate.

Some of the interviewed people also linked climate change to the controversial Fast Track Land Reform Programme which coincided with drought and famine in the area. To them, climate change is directly linked to the political and economic crisis that has gripped the country since 2000. The implication is that when there are political, social and economic problems in a country, agro-pastoralists tend to link them to climate change (Mubaya 2011a). In addition, the cultural context and spiritual world-view play a critical role in shaping farmers' perceptions and attitudes, a factor which may cloud farmers' consciousness of the negative effects of human activities on the earth systems (Mubaya 2011b). Before the year 2000, the informants noted that rainfall patterns were 'normal' compared to the current situation where they rarely experience river flooding during the rainy season. When asked about the most recent natural disaster which was attributable to climate change and variability, informants identified Cyclone Eline. According to most of the interviewed people, the impact of Cyclone Eline which occurred in February 2000 is still visible in the area. The natural disaster left a trail of destruction especially potholes along the Harare-Beitbridge highway.

Key informants indicated that droughts have become a perennial occurrence in Chivi District. They also highlighted that although they have not experienced floods since the infamous Cyclone Eline, low rainfall patterns have impacted negatively on them in many ways - crops suffering from moisture stress, crops failure, water shortages, human and livestock diseases, among others. The interviewed people from the four villages concurred that before the turn of the century it was easy to predict the coming season; and the seasons were distinct but now the rains have become increasingly unpredictable. Moreover, they also highlighted that now they are experiencing shorter rainy seasons than before. According to them, in the past, rainy seasons started around 15th October but now it only starts raining around mid-November. These findings are consistent with the gloomy picture painted in the literature on climate variability and change in Africa, in general, and southern Africa, in particular (Mubaya 2011a), where the late start to the rainy season, prolonged mid-season droughts, and shorter growing seasons have become the norm. It is noteworthy to highlight that anthropogenic factors raised by the 2007 Inter-governmental Panel on Climate Change (IPCC) fourth assessment reports as driving climate change were not mentioned by the study participants. This is consistent with research done on other human tragedies such as HIV and AIDS, where interviewed people identified exogenous rather than endogenous factors as responsible for their misery.

Thick descriptions on climatic changes in Chivi District came from life history interviewees who expressed concern on the deteriorating bio-diversity in the area. Eight life history interviews (two in each village) revealed that weather patterns (rainfall, wind, temperatures and humidity) have changed. Interviewees indicated that visible signs such as the failure of certain wild fruits to bloom and produce fruits during the season were enough evidence. They also complained of unusually high temperatures especially in October, November and December. Besides high temperatures, people in Chivi have started experiencing shorter and warmer winters than before. When asked how they assessed weather patterns, interviewed people revealed that they rely on indigenous weather forecasting techniques. Most of the life history interviewees had lived in their respective villages for an average of 35 years which enabled them to recall watershed moments in climate change. For instance, a 79 year-old woman from Chikanda Village aptly captured the changes in climate over the past couple of years as follows:

> In the 1980s, we used to have normal rains and bumper harvest. Although Chivi has always been a dry place, things were better. I can recall many families in this village had a lot of livestock. Our grazing lands were protected areas but now as you can see, those areas have been turned into residential areas to cater for population growth. Our livestock are now being forced to graze on mountains without lush green grass. On the other hand, population growth has coincided with low rainfall patterns and hot temperatures, which has seriously affected our traditional agricultural practices. Fewer livestock have complicated our efforts to harvest manure which we usually use to improve soil fertility. Wild fruits which we used to harvest all-year-round have also disappeared from our forests. This is ample evidence that all is not well. All our major rivers are empty. Droughts have become the order of the day. So, for me climate change is upon us. We cannot pretend that it is business as usual.

The agricultural extension officer in the area also weighed in with his opinion:

> We do not need a rocket scientist to tell us that we are in the middle of the harshest summer in living memory. This has been scientifically confirmed by the Meteorological Department, which said these high temperatures have broken the records set in 1961. If you have been living in this community over the past few years, you might have noticed that our rainy season has shifted from October to early December.

His views resonated with opinions expressed by most participants of the FGDs. Although interviewees noted that it was difficult for them to quantify and qualify the degree to which climate is changing, certain signs were evident to the ordinary villagers. Study participants also indicated that trees like baobab which are known to be drought resistant are also being affected. Furthermore, people in Chivi District used to hunt animals to supplement their diet, but this is no longer the case as there is nothing to hunt. They pointed out that a lot of wild fruits were no longer flowering and germinating in season. One elderly woman had this to say:

> We are no longer able to brew *mukumbi* or *amarula* because the raw materials which we get from a wild fruit popularly known as *mupfura* haven't been germinating on time over the past couple of years. This is the time when we would brew *mukumbi* in this area but as you can see at this village meeting, no one is sloshed up here. We used to make money from selling the illicit brew but things have changed. We are also travelling long distances in search of green pastures. Nature has imposed its own sanctions on us. Who are we my son to argue with the creator?

The above interview extract shows that even wild fruits have not been spared by the debilitating impact of climate change in Chivi District. Even pastures for livestock have been severely affected forcing villagers to travel to faraway places like Nyajena and Tugwi in search of pastures.

Men in the focus group discussions indicated that they had lost a number of livestock due to increased cases of theft, drought and animal diseases. They bemoaned the lack of veterinary medicine in the area. Dipping chemicals were also in short supply, thereby exposing their livestock to various diseases.

Besides the shortage of wild fruits and pastures due to the receding water table in the area, interviewees indicated that the impact of prolonged droughts had dried up natural water points such as streams and rivers. Findings from the women's focus group discussion indicated that it was now difficult to irrigate their gardens at a nearby dam because water levels made it difficult to use the canal situated in the middle of the communal nutritional garden. For instance, participants from Zivuku and Machanja bemoaned the shortage of water. Most of them have resorted to drawing water from Chendebvu Dam by using scotch carts and wheelbarrows. The duty of fetching water was noted as hindering women from engaging in productive household chores, because they spent a good part of their time walking long distances to fetch water for domestic consumption.

Livestock were found to be more vulnerable to foot-and-mouth diseases in the area than ever before. Weather conditions were described as extremely hot.

Findings from life history interviewees and women's focus group discussions demonstrated that to some extent, the participants are well versed on the occurrence of climate variability and change in the area. Their views are not so detached from what is found in the climate change literature especially with regard to high temperatures and increased cases of drought and famine (Omolo 2011; Ongoro and Ogara 2012). Increasing temperatures and variations in rainfall patterns over time, coupled with frequent mid-seasonal droughts and cyclones are clear evidence that the climate has changed in Chivi. Overall, the findings reflect agro-pastoralists' perceptions of climate change are based on their traditional stocks of knowledge.

Gender-Differentiated Impacts of Climate Change and Variability

The issue of gender-differentiated impacts of climate change attracted a huge interest from the participants in the four FGDs held in Chivi District. Although the men and women participating in the study noted that climate variability is impacting on everybody who is poor and depends on activities and resources that are threatened by climate, gender-differentiated impacts manifested themselves in terms of roles and responsibilities. According to the traditional gender-based division of labour, women are responsible for child-bearing, caring for children, the sick, and the elderly (home-based care); subsistence farming, weeding the fields; tending family nutritional gardens; harvesting and storage of food crops; food preparation; collection of firewood; fetching water from the river and borehole; collecting thatching grass; and brewing beer for ritual and commercial purposes. On the other hand, men are expected to provide for the family financially (bread-winners), protect the family (safety), plough the fields using draught power, engage in commercial farming, herd and milk livestock, treat sick livestock, attend communal and village meetings, build and thatch houses, build kraals and fence houses and fields.

In the traditional gender-based division of labour spelt out above, duties considered manual and requiring a lot of energy are set aside as masculine while the opposite are seen as feminine. However, it is important to point out that although the private and public sphere is reflected in the duties allocation, there is an emerging trend towards the blurring of the two spheres. Nevertheless, this does not obfuscate the unequal distribution of roles and responsibilities in rural Chivi. The blurring of the private and public spheres has been accentuated by the feminisation of rural areas and mass migration in Chivi District. For instance, the herding of livestock is increasingly becoming feminised. Both men and women concurred that their roles and responsibilities are in a state of flux. Baba Morgan, a retired mine worker, explained the gender-differentiated impacts of climate change and variability on agro-pastoral communities as follows:

> Since the start of recurrent droughts associated with climate change, we have seen more and more women in this community doing household chores which are traditionally considered masculine. Fetching firewood with scotch carts, pushing wheel barrows full of water containers and ploughing the fields are now done by women as most men have left the community in search of greener pastures. Therefore, it is clear that climate change has further burdened women who are still expected to cook, clean their houses and attend village meetings on behalf of their absent husbands.

Mai Chikoto, a 70 year old who had lived in Chivi since birth, echoed Baba Morgan's statement when she narrated how women are almost becoming *de-facto* heads of households as more and more able-bodied men are leaving the area in search of employment opportunities:

> We are seeing more women acting as heads of households in the absence of their husbands. Climate change came at a time when women were already taking up extra workloads relinquished by absent men. As you can see at this village meeting, we have very few able-bodied men, so who do you think ploughs the fields? These women you see here are an industrious lot. Climate change has certainly colluded with other factors to worsen the burden thrust upon the shoulders of women while at the same time reversing the traditional gender-based division of labour.

The above interview excerpts demonstrate that climate change is combining with other stressors, thereby creating a complex vulnerability trap for women. It must be noted that although women are becoming *de-facto* heads of households, this has a knock-on effect on other duties such as childcare, home-based care and feeding of families.

However, with the blurring of the private and public spheres, young men are also actively involved in collecting water and firewood. These are all activities that were previously dominated by women and girls. A shift in gender roles is not only unique to these communities but has also been observed in other communities (WEDO 2007; Ongoro and Ogara 2012).

Most female participants concurred that their traditionally defined roles as cooks put too much pressure on them. Some women even revealed that they are suffering from high blood pressure and stress especially when they fail to live up to their socially defined roles.

There was unanimous agreement among most FGDs participants that women are carrying extra-workloads, which predisposes them to working longer hours compared to men. Men were reported to have more leisure time as compared to their women counterparts. On the other hand, interviews with men indicated that they were also aware of the burden and heavy workload that women endure while trying to secure food to sustain the household, with some of them expressing frustration with crop failure induced by climate change and variability. Both men and women reported that feelings of helplessness and stress were compounded by multiple stressors which made it difficult for them to devise sustainable survival options. Psychological and physical effects of climate change are coalescing, thus creating a vicious cycle for women (Babugura 2010).

Traditional and New Coping and Adaptation Mechanisms

Historically, besides livestock, major food crops in Chivi District included cereals such as maize, wheat, sorghum, millet and vegetables; while the major cash crops have been cotton, tobacco and sorghum. However, this has dramatically changed as yields have continued to dwindle and livestock have succumbed to numerous disease outbreaks and drought.

The adaptability of agro-pastoralism, which depends on water availability and pastureland productivity, has now been negatively impacted by unpredictable climate change, and enhanced environmental degradation. Most of the agro-pastoralist families in Chivi District are responding to the adverse impacts of climate change by deploying a variety of survival strategies.

It is clear that agro-pastoralists possess human agency (Long 1992), which is the capacity of actors to process their social experiences, and learn from others' experiences and to act upon them. Informants also noted that, given the economic challenges which the country is facing, rearing livestock for sale and planting cash crops were no longer sustainable as the prices were very low. For instance, selling a bullock to local butcheries or meat traders fetched as little as US$300 which was not enough to send a child to a decent boarding school. In the four FGDs with mostly women, most participants acknowledged switching to other crops and varieties that are drought resistant (such as SeedCo and Pannar seed varieties); engaging in multiple cropping and inter-cropping practices; investing in alternative community gardens; and irrigation facilities.

Overall, women in Chivi District indicated that they have put in motion a number of survival strategies. These include selling illicit beer, tending nutritional gardens, cross-border trading and roadside vending.

Young women, on the other hand, have limited mobility due to the powerful idiom of domesticity (Mate 2002). Some of them have been forced into early marriages, commercial sex work (mostly in beer halls at Ngundu Growth Point); while others engage in street vending along the Masvingo-Beitbridge Highway.

In contrast, old and young men have resorted to buying and selling fuel along the Harare to Beitbridge Highway, venturing in sculpture businesses which have led to severe deforestation as they cut down *mukamba* trees to continue producing artifacts.

Young men have resorted to cross-border jumping (into South Africa, Mozambique and Botswana), rural to urban migration, venturing into roadside fuel service stations, buying and selling livestock, gold panning and foreign currency dealing.

Through this immersion in the county's burgeoning speculative informal economy, young men demonstrate that "they are self-aware actors in charge of their lives and destinies rather than defenceless victims of circumstances" (Giddens 1986).

It is noteworthy to point out that some of the survival strategies deployed by young men in Chivi District are dependent on the natural environment which amplifies anthropogenic factors driving the latest trend of climate change. For instance, gold panning in Runde River is contributing to soil erosion and the siltation of major surface water bodies (Dhliwayo 2007). Cutting down trees in search of firewood for sale and wood carvings has spawned deforestation in the area.

As already noted, survival strategies are being deployed at the individual and household level. It is noteworthy to highlight that child- and grandmother-headed households were identified as barely surviving. School drop-out rates amongst these households were described as high compared to those with either one or both parents. Early marriages were also more pronounced amongst female and child-headed households as a coping mechanism.

Informants pointed out that survival strategies adopted at the household level in Chivi District included cutting down on food consumption, making dietary adaptations, switching to drought resistant crops, use of traditional medicine to cure livestock, distress sale of livestock, setting up cooperative schemes (community nutritional gardens), being part of group savings (merry-go-rounds), organising systems of group labour (*humwe* or *nhimbe*) for assisting one another in weeding crops and herding livestock and engaging in income-generating projects. Clearly, a new culture is thus emerging in Chivi District that emphasises resilience, innovation and group collectiveness. A good example of this is the system of group labour (through the Shona system of *humwe* or *nhimbe*) and setting up group savings (*maRounds*) by women in the four villages studied. Social networks based on kinship, extended family ties and religious affiliation play an instrumental role in times of need, such as periods of drought and famine.

These popular coping strategies in Chivi District demonstrate agro-pastoralists as "social actors are able to monitor their own thoughts and activities as well as their physical and social contexts" (Giddens 1986). In addition, findings from this study concur with Mitchell *et al.*'s (2007) observation that Nepalese women have resorted to seeking employment, secretly saving part of their earnings, distress sale of livestock and setting up of community-based self-help groups and networks, and group savings, or systems of group labour .

Major Alternative Livelihood Strategies among Agro-pastoralists in Chivi District

The study has found out that agro-pastoralists in Chivi have come up with various survival strategies to cope with their fragile existence. The various modes of or mechanisms for survival adopted by agro-pastoralists in their bid to cope with the vagaries of climate change include both positive and negative ones. The positive ones mainly include the entrepreneurial skills that individuals and households have developed in response to their precarious environmental and economic situations. These include taking part in food for work programmes, food aid, migration, street vending and cross-border trade. On the other hand, negative coping mechanisms include prostitution and cattle rustling.

Food-for-Work Programmes and Food Hampers from NGOs

In the four FGDs, it was revealed that nutritional food-for-work programmes and food aid are important alternative livelihood strategies in Chivi District.

Food-for-work programmes run by government agencies such as the District Development Fund and non-governmental organisations were also identified as essential.

FGD participants indicated that food aid, especially distributed by non-governmental organisations such as Care Zimbabwe, World Food Programme and Dutch-Care, have helped to turn the tide against starvation. They revealed that households classified as 'food insecure and vulnerable' were entitled to food hampers which comprised cooking oil, beans, barley and maize meal. Mai Agnes summed up her over-reliance on food aid in the following interview excerpt:

> Without Care Zimbabwe, I don't know how I could have managed to survive because I don't have a husband and all my kids have perished due to HIV and AIDS. I am too old to look for employment. Non-governmental organisations have become more like my shelter in times of need. As a widow, food aid has become my hope in this area. For the past ten years, I have failed to harvest more than four bags of maize and as you can see all my cattle died too as a result of foot-and-mouth diseases. I rely on others to plant my one hectare plot. Although we get food hampers after a long time, we have learnt to use sparingly so that we can at least have one meal per day.

The above statement captures the struggles facing widows and orphans in the area. It was also revealed that those households without a single person in the Diaspora were more likely to be vulnerable to lack of food and other basic needs.

In-depth interviews with key informants revealed that primary schools in the area have benefited tremendously from food hampers from non-governmental organisations.

The elderly and orphans in the community were also identified as beneficiaries of food hampers. In the absence of government social grants, it must be emphasised that non-governmental organisations have filled the vacuum. It is important to note that at the time of this fieldwork, at least 29 non-governmental organisations operating in Masvingo Province were banned by Governor Titus Maluleke for operating without proper registration.

Migration: Cross-Border Jumping

Climate-related stressors, combined with ecosystem change, have the potential to drive migration or prompt national governments to plan for the relocation and resettlement of affected populations (Warner *et al.* 2008). It is believed that as some "environments become inhospitable, these people move elsewhere where their locally specific knowledge may no longer apply" (*Ibid.*). However, male-biased migration processes have left women to cope with, adapt to and mitigate the impacts of climate change in Chivi District. Given the impact of proletarianisation processes on gender relations, women are considered immobile based on the jaundiced application of the Victorian ideology and misunderstanding of the African customary family relations (Mate 2002). Men, therefore, are seen as 'bread-winners'. Interviews conducted with both men and women in Chivi confirmed that many families in the area had one or more children in the Diaspora. This also confirms Giddens' (1986) insight that human beings are able to rationalise and develop routines that enable them to deal efficiently with their social lives. In the case of agro-pastoralists, migration has become an integral livelihood strategy. Baba Tsvakai, a retired teacher from Zivuku Village explained the role of cross-border jumping as family survival strategy thus:

> Without cross-border jumpers, this community could have been wiped out by hunger especially in 2008. Most of our kids working in South Africa come to our rescue by periodically sending remittances in the form of cash and groceries. Due to their hard work, we have managed to build two modern houses and also bought solar panels. Even though our livestock population has been depleted by famine and diseases, we are food secure as long as our children in the Diaspora continue to send us groceries.

However, male migration has increased the women's workload, as they are left behind to manage the households by themselves in addition to their usual tasks. Mai Tachiona, a resident of Vhudzi Village, also confirmed that recurrent droughts blamed on climate change, coupled with limited employment opportunities for school graduates, had fuelled the process of mass migration by young people to neighbouring countries. In her words:

> Before the dawn of the Zimbabwean crisis, very few people migrated to South Africa in search of employment. In this community, mass migration was spurred on by climate change especially recurrent droughts. As from 2001, cross-border jumping has become the most popular survival strategy for the youth. Social networks of border jumpers have sprung up which have helped new ones to cross safely as well as facilitated the transfer of remittances and groceries back home. However, some kids from this community have also drowned in Limpopo River as they tried to cross into South Africa illegally while others have gone forever.

Clearly, remittances have become an important aspect of Chivi residents' and migrants' social security system. Investments made at home through remittances are not only relied on in times of need, but are also meant to sustain migrants and their families when they eventually return to Zimbabwe (Muzondidya 2008). Focusing on migrants from the Matebeleland South districts of Bulilima, Mangwe, Gwanda, Beitbridge and Matobo, Maphosa (2004, 16-17) concludes that remittances contribute significantly to the improvement of the livelihoods of the receiving households. According to Maphosa (*Ibid.*), besides investing in houses, solar power and boreholes, these migrants invest in bicycles and scotch carts which are the major forms of transport in these rural areas.

Cross-Border Trade

Results from FGDs also confirm that women play an important role in supporting households and communities to cope with and adapt to climate variability. For instance, Mai Batanai, an agro-pastoralist woman from Chivi, narrated how cross-border trading had become pivotal for the survival of her family:

> I started engaging in cross-border trading in 2002 and it has helped my family a lot because we have not harvested anything significant in this community for a long time. Our livestock have also died due to the shortage of pastures and water. So, because of perennial hunger, I decided to become a cross-border trader, while continuing to engage in agriculture during the rainy season. This has changed my family fortunes a lot as this means I have access to cash and cheaper groceries. I don't know what would have become of my family had I not started this business.

It is clear from the foregoing that women are entrepreneurial and resilient actors in the face of the debilitating impact of climate change. Similar sentiments were echoed by Mai Chokuda, a villager from Machanja. She explained how cross-border trading was helping her cope with the environmental and economic crisis as follows:

> Since I started travelling to South Africa as a cross-border trader in 2005, things have become better. My husband is a retired teacher whose pension has been rendered worthless due to the hyper-inflation during the peak of the Zimbabwean crisis. As I am talking to you now, I have managed to furnish our four-roomed house, clothe our children and send the eldest daughter to university. Before I began this business venture, I used to rely on food aid and support from the extended family members.

Mai Batanai and Mai Chokuda's accounts reveal that cross-border trading has become an integral family survival strategy. This also shows that women are indeed "powerful agents for development and change" (Enarson 2008) in the context of climate change in Chivi District. Women indicated that they travel to places as far as Mesina (South Africa) and Francistown (Botswana), where they buy basic commodities, electrical gadgets, solar panels, clothes and blankets for sale. They also explained that they usually travel on Friday night when school kids are back home and return on Saturday. Through cross-border trade, most interviewed people indicated that they have managed to financially empower themselves and stabilised their family circumstances. Far from being *tabula rasa* or blank slates, agro-pastoralists in Chivi District are using their "practical consciousness or stocks of knowledge" (Giddens 1986) to cope with the impact of climate change.

Negative Survival Strategies

Finding themselves in precarious environmental and economic positions, some villagers in Chivi District have found themselves resorting to capricious survival strategies, involving criminality and corruption.

It was also found out that although women begin by deploying positive survival strategies, given their vulnerable situation, most ended up engaging in commercial sex work with truck drivers.

Women who work as roadside vendors were singled out as highly vulnerable to sexual advances by truck drivers plying the Harare-Beitbridge Highway. Interviews with key informants also confirmed this phenomenon. Findings from women's focus group discussions indicated that prostitution was popular amongst young women. Thriving commercial sex business was described as rampant at Ngundu Growth Point, a business centre which is approximately 90km from Masvingo. Traditional leaders also noted that the growth point was spawning moral decadence in their areas as young men and women were camping for days before returning home. This was confirmed by one key informant who noted that school drop-out rates, early marriages among girls and prostitution in the area were alarming. The following excerpt is illustrative:

> This area has witnessed a social crisis in recent years. Most boys are dropping out of secondary schools as they are lured by the prospects of earning Rands in South Africa. For them, border jumping is a passport to a better life compared to finishing school and wallow in poverty forever. They simply need US$10 to get to Beitbridge and then jump the border into South Africa where they work as agricultural labourers on plantations. On the other hand, the girls who remain at school rarely finish because once these cross-border jumpers return home armed with gadgets of modernity and flashing a few Rands, they are lured into 'one night stands' which lead to unwanted pregnancies and early marriages. This has become the unwritten script of most youth in this area since 2000.

Migration has become a double-edged sword in Chivi District. On the one hand, it has brought the much-needed remittances, gadgets of modernity (solar panels, television sets, bicycles, cell phones, transistor radio sets, power generators and cars) and youth employment. On the other hand, it has contributed to moral decadence, alcohol abuse, unwanted pregnancies, school drop-outs and high divorce rates. It is arguable that climate change has coalesced with economic difficulties obtaining in the country to the point of emasculating men from their traditional bread-winner responsibilities.

Furthermore, the laying off of workers during the 'Zimbabwe crisis' era coupled with crops failure was blamed for fostering a 'masculinity crisis' which has forced some men to engage in multiple concurrent relationships. Such risky sexual behaviour has put married women at the risk of contracting HIV/AIDS especially in the context of marriage. Besides, putting the lives of men at risk of contracting HIV and other sexually transmitted diseases, these negative survivalist strategies also endanger the lives of women and children.

Also accompanying negative survival strategies are cases of gender-based violence mostly perpetrated by men under the influence of alcohol.

In summary, it is important to point out that, although most households are trying to cope with climate change, some villagers in Chivi District are barely surviving because of limited other livelihood options. Thus, far from romanticising and generalising the agency of agro-pastoralists in Chivi District, it can be concluded that while some women are deploying positive survival strategies others are falling by the wayside. These findings are congruent with other studies on gender and climate change in pastoral communities (Omolo 2011; Ongoro and Ogara 2012), which conclude that women, due to the socially constructed gender roles, are more vulnerable to the impacts as compared to their male counterparts. It has also shown that far from resigning their lives to the fate of the environment, agro-pastoralists in Chivi District are resorting to both positive and negative survival strategies.

Conclusion

From the findings discussed above, it is clear that based on the common-sense knowledge of agro-pastoralists in Chivi District, climate variability is in full swing. Despite acknowledging the occurrence of climate variability, agro-pastoralists were not able to distinguish it from climate change. They assume that indicators such as poor agricultural yields, rising temperatures and a spike in social ills can only be explained within the realm of socio-cultural and spiritual factors. Agro-pastoralists in Chivi District, however, are resorting to survivalist responses at the individual and household levels in order to remain afloat. These include: cross-border jumping, cross-border trade, foreign currency dealing, prostitution, roadside fuel vending, sculpture-making, food-for-work programmes, walking long distances in search of livestock pastures, experimenting with different seed varieties, marrying off girls to cross-border jumpers and reliance on food aid distributed by non-governmental organisations. Moreover, due to high levels of migration among the youth and able-bodied men who are crossing into South Africa in search of employment, climate change and other multiple stressors have contributed immensely to the reconfiguration of gender-based division of labour, feminisation of poverty, feminisation of rural areas and increased cases of child labour.

In order to deal with a multi-faceted environmental and economic crisis confronting agro-pastoralists in Chivi District, youth unemployment in rural areas needs to be tackled through the promotion of income-generating projects and rural electrification projects which create employment opportunities. There is a need to improve the effectiveness of the Basic Education Assistance Module (BEAM) through the revitalisation of the means-test and better beneficiary targeting strategies especially in rural areas.

Traditional leaders and residents of Chivi need to revive the practice of *Zunde RaMambo* (chief's granary), a traditional social security arrangement designed as a contingency of drought or famine. This practice has the ability to help residents in times of food insecurity and mitigate growing levels of malnutrition in rural areas. In order to make the practice of *Zunde RaMambo* effective as community response to food insecurity, there is a need for the government to allocate land and inputs to villagers and their traditional leaders.

Given the over-dependence of the people in Chivi on food aid, there is an urgent need for the government to lift its ban on 29 non-governmental organisations operating in Masvingo which are accused of operating without registration.

Government ought to continue with the national grain loan scheme.

There is also a need to establish the pass-on heifer scheme in order to restock the national herd of livestock in communal areas such as Chivi District.

Veterinary services in rural areas need to be improved and monitored regularly.

At the policy level, the study also highlights the need to mainstream gender with the policies and legal frameworks that anchor the adaptation and mitigation of climate change in Zimbabwe such that there is full participation of both men and women for burden-sharing and sustainable development.

It is critical for the government to design integrated social welfare and climate change policies which foster adaptive capacity at the community level.

References

Anderson. C. 2002. Gender matters: Implications for climate variability and climate change and for disaster management in the Pacific Islands. Available from: http://www.gencc.interconnection.org. (accessed on 23rd October, 2011)

Ariyabandu, M., and Wickramasinghe, M. 2003. *Gender dimensions in disaster management: A guide for South Asia*. Colombo: South Asia Publications.

Babbie, E. and Mouton, J. 2001. *The practice of social research*. Oxford: Oxford University Press.

Babugura, A. 2010. *Gender and climate change: South Africa case study*. Cape Town: Heinrich Boell Foundation.

Birch, I. and R. Grahn. 2007. Pastoralism – Managing multiple stressors and the threat of climate variability and change. In: United Nations Development Programme, *Human development report 2007/2008: fighting climate change: human solidarity in a divided world*. UNDP: Dhaka.

Bollig, M. and A. Schulte. 1999. Environmental change and pastoral perceptions: Degradation and indigenous knowledge in two African pastoral communities. *Human Ecology*, 27(3): 493-514.

BRIDGE. 2008. Gender and climate change: Mapping the lLinkages. BRIDGE Institute of Development Studies (IDS), UK. Available from: http://www.bridge.ids.ac.uk. (accessed on 20th November, 2011)

Brody, A., Demetriades, J., and Esplen, E. 2008. *Gender and climate change: Mapping the linkages*. London: Institute of Development Studies (IDS).

Carvajal . Y., M. Quintero, and M. Garc´ıa. 2008. Women's role in adapting to climate change and variability. *Advances in Geosciences*, 14: 277-280.

Dankelman, I. 2002. Climate change: Learning from gender analysis and women's experience of organising for sustainable development. *Gender and Development*, 10 (2): 21-29.

Denton, F. 2001. Climate change, gender and poverty – Academic babble or realpolitik? *Point de Vue*, 14, 1-2.

——.2002. Climate change vulnerability, impacts, and adaptation: Why does gender matter? *Gender and Development* Vol. 10 (2): 10 - 20.

Dhliwayo, M. 2007. Human rights and climate change in Chivi. Harare: Zimbabwe Environmental Law Association. Available from <http://www.ciel.org/publications/climate/change/in/chivi. (accessed on 21st September, 2011)

Enarson, E. 2008. *Gender and natural disasters*. IPCRR Working Paper No. 1. Geneva: International Labour Organisation.

Galvin, K.A.; P.K. Thornton; R.B. Boone, and J. Sunderland. 2004. Climate variability and impacts on east African livestock herders: The Maasai of Ngorongoro Conservation Area, Tanzania. *African Journal of Range and Forage Science*, 21 (3), 1-2.

Gandure, S. 2005. Coping with and adapting to drought in Zimbabwe. PhD diss. Johannesburg: Faculty of Science, University of Witwatersrand.

Giddens, A. 1986. *The constitution of society outline of the Theory of Structuration,* Berkeley: University of California Press.

Hartmann, I. and J. A. Sugulle. 2009. The impact of climate change on pastoral societies of Somaliland. *Candlelight for health, education & environment report*, November 2009.

Intergovernmental Panel on Climate Change (IPCC). 2001. Summary for policy makers. In *Climate change 2001: Impacts, adaptation and vulnerability*. Intergovernmental Panel on Climate Change. Cambridge: Cambridge University Press.

——.2007. Climate change 2007 – impacts, adaptation and vulnerability. Contribution of Working Group II to the Fourth Assessment Report of the IPCC. Cambridge: Cambridge University Press, UK. Available from: http://www.ipcc.ch/activity/wg2outlines.pdf. (accessed on 11th November, 2011)

Jost, C.C. 2002. *Facilitating the survival of African pastoralism in the face of climate change*. Boston: M.A. Fletcher School of Law and Diplomacy, Tufts University.

Kirbride, M. and R. Grahn. 2008. Survival for the fittest: Pastoralism and climate change in East Africa. Oxfam International Briefing Paper, August 2008. Available from http://www. ciaonet.org/pbei/oxfam/0003433/f_0003433_2544.pdf (accessed on 10th February, 2012)

Long, N. 1992. From paradigm lost to paradigm regained: The case of actor-oriented sociology of development. In N. Long and A. Long (eds.), *Battlefields of knowledge: the interlocking of theory and practice in social research and development*. New York: Routledge.

Maphosa, F. 2004. *The impact of remittances from Zimbabweans working in South Africa on rural livelihoods in the Southern Districts of Zimbabwe*. Dakar: CODESRIA.

Marshall, M. N. 1996. Sampling for qualitative research. *Journal of Medical Education* 45 (9): 700-705.

Mate, R. 2002. Creating and celebrating "African-ness"– juggling with men and women's interests in family life: The status of customary law as a measure of political commitment to gender equality in Zimbabwe. Paper presented at CODESRIA's 10th General Assembly on 'Africa in the New Millennium", 8th – 12th December, Kampala.

Meier, P., D. Bond, and J. Bond. 2007. Environmental influences on pastoral conflict in the Horn of Africa. *Political Geography*, 26 (6): 716-735.

Mitchell, T.; T. Tanner, and K. Lussier. 2007. *We know what we need! South Asian women speak out on climate change adaptation*. London: Action Aid International and the Institute of Development Studies (IDS).

Mubaya, C. P. 2011a. Climate variability and change or multiple stressors? Farmer perceptions regarding threats to livelihoods in Zimbabwe and Zambia. Paper read at Afrika Adapt Symposium, 9-11 March, Addis Ababa.

–––.2011b. Carving a niche for the social sciences in multi-disciplinary research on climate change adaptation and agriculture in Southern Africa. Paper read at the ISSC-CIPSH General Assembly Joint Symposium, 1st May, Nagoya.

Muzondidya, J. 2008. Majoni-joni: Survival strategies among Zimbabwean migrants in South Africa. Paper presented at the International Conference on the Political Economies of Displacement in Zimbabwe, 9-11 June, at Witswatersrand University, Johannesburg.

Mworia, J.K. and J.I. Kinyamario. 2008. Traditional strategies used by pastoralists to cope with La Nina induced drought in Kajiado, Kenya. *Africa Journal of Environmental Science and Technology*, 2 (1): 23-56.

Nasreen, M. 2000. Coping mechanisms of rural women in Bangladesh during Floods: Gender perspective. In: *Disaster: Issues and gender perspectives*. Conference Proceedings, University of Dhaka, June 23 & 24, 2000.

O'Brien, K. 2007. Commentary to the paper of Úrsula Oswald Spring, 'Climate change: A gender perspective on human and state security approaches to global security'. Paper presented at International Women Leaders Global Security Summit, 15-17 November, New York.

Omolo, N. A. 2011. Gender and climate change-induced conflict in pastoral communities: Case study of Turkana in north-western Kenya. Paper presented at Afrika Adapt Symposium, 9-11 March, Addis Ababa.

Ongoro, E. B., and W. Ogara. 2012. Impact of climate change and gender roles in community adaptation: A case study of pastoralists in Samburu East District, Kenya. *International Journal of Biodiversity and Conservation,* 4(2): 78-89.

Parikh, J. 2007. *Gender and climate change framework for analysis, policy and action*. New Delhi: UNDP India and Integrated Research and Action for Development.

Petrie, B. 2010. *Gender and climate change: Regional report*. Cape Town: Heinrich Boell Foundation.

Pratt, D.J., F. Le Gall., and C. De Haan. 1997. *Investing in pastoralism: Sustainable natural resource in arid Africa and Middle East*. World Bank Technical Paper 365, Washington, D.C.: World Bank.

Swift, J. 2003. *Pastoralism and mobility in the drylands: The global drylands imperative*. New York: UNDP.

Taylor, F. and P. Bogdan. 1984. *Social science research*. London: Blackwell Publishers.

Tigere, C. 2011. *Climate change and adaptation studies in Zimbabwe*. Cape Town: Heinrich Boell Foundation.

Warner, K.; T. Afifi; O. Dun; M. Stal. and S. Schmidl. 2008. *Human security, climate change and environmentally induced migration*. Bonn: United Nations University Institute for Environment and Human Security (UNUEHS).

Warren, R. 2006. Impacts of global climate change at different annual mean global temperature increases. In Schellnhuber et al. (eds.). 2006. *Avoiding dangerous climate change*. London: Cambridge University Press.

Women's Environment and Development Organisation (WEDO). 2003. *Common ground, women's access to natural resources and the United Nations Millennium Development Goals.* New York: WEDO.

—–.2007. *Changing the climate: Why women's perspectives matter*. New York: WEDO.

—–.2008. *Gender, climate change and human security: Lessons from Bangladesh, Ghana and Senegal*. Report Prepared for ELIAMEP, New York: WEDO.

Chapter 8

Effects of Climate Change and Variability on Pastoral Communities: Coping and Adaptive Strategies for Women and Men in Uganda

Consolata Kabonesa[*] and Fredrick Immanuel Kindi

Abstract

This study examines the difference in coping and adaptive strategies to climate change between men and women in pastoral communities and how such strategies are influencing their livelihoods. Methodologically, the study has used qualitative primary and secondary data that was gathered through the Rapid Appraisal Technique, using focus group discussions as well as in-depth interviews with key informants. Findings support the existing claims that gender influences men's and women's coping and adaptive strategies. Participants noted a number of coping and adaptive strategies ranging from, but not limited to, sale or slaughter of cattle; diversification by engaging in other income-generating activities such as charcoal-burning, business/trade; supplementing livestock with agriculture, as well as temporary migration to river banks and lake shores as well as to neighbouring districts to graze and water their animals. Other strategies have included adoption of improved breeds, tree planting, sensitising and training communities about the need to preserve the environment, and construction of

[*]ckabonesa@ss.mak.ac.ug; kindifred@ss.mak.ac.ug

valley dams. For women in particular, they have had low coping and adaptive capacity due to inequality in access to and control over resources such as land and livestock, limited opportunities for education and decision-making. Climate change and variability is increasing women's workload as men migrate in search of alternative income-generating activities although, in the long term, men's absence permits women the autonomy and independence to assume household leadership, hence enhancing their agency. In general, however, there is a need for urgent interventions to ensure that autonomous adaptations do not exacerbate the environment, while planned adaptations have to be sensitive to men's and women's socio-economic and cultural conditions in influencing their coping and adaptive capacities.

Introduction

Background

In the recent past, changes in climatic conditions and environmental degradation across the globe have captured international attention. These climatic changes constitute a serious threat to human survival. On the one hand, human activities, especially those that disregard the environment, are seen as degrading natural resources such as water, forests, fertility of soils, and in the process, inducing climatic changes and variations. Such human activities include the burning of fossil fuels from cars and industrial processes, changes in land-use, like deforestation, and the release of greenhouse gases into the atmosphere (SACAU 2009). On the other hand, these climatic changes are viewed as normal processes and are characterised by high temperatures, flooding, long spells of dry seasons, and unpredictable rains. The combined effects of human activities and the changes in climatic conditions are leading to much more complex and potentially disastrous situations, especially among agricultural and pastoral communities in different parts of the globe. As such, adapting to climate change has become a major policy priority for many countries in sub-Saharan Africa (SSA), as well as their development partners.

Within the context of pastoral communities, women and men contribute a great deal of labour and other pastoral-related activities that make significant input to the Gross National Product of countries where they live. The pastoralists provide meat, hides and skins, and because they operate mostly in drylands they are custodians of dryland environments, and provide services through good rangeland management, including bio-diversity conservation and wildlife tourism (Oxfam 2008). Despite their contribution, pastoral communities across a number of regions on the globe are seen as poor and resistant to modern animal husbandry methods. For example, in some East African countries such as Uganda, 64 per cent of the pastoralist population live below the poverty line compared to 38per cent nationally (Oxfam 2008). With the prevailing changes in the climatic conditions, pastoral communities now face a number of challenges that hinder their lifestyle and are compelled to adapt to climatic changes, as pasture and water for their animals are becoming scarce. Although several studies have attempted to present the threat of climatic change and environmental degradation in gender-neutral terms, a rising number of researchers are advocating studies especially in developing countries that examine climate change and variability from a gender perspective. Adequate studies are yet to be conducted on how climate change and variability are impacting on women and men, their adaptation, and how this affects their livelihoods. This chapter, therefore, responds to this inadequacy.

Problem Statement

Whereas plenty of studies have focused more on adaptation to climate change among agricultural communities (Parikh 2007; SACAU 2009; Skinner 2011), research is yet to be conducted among pastoral communities, especially from a gender perspective. Studies in pastoral communities report how changes and variability in climate are impacting on pastoralists (Oxfam 2008; Djoudi and Brockhaus 2011). For instance, prolonged droughts are leading to conflicts as pastoralists search for water and pasture, deaths of livestock, with the surviving ones becoming thin, weak and less productive. Consequently, such pastoralists have to devise ways of coping and adapting. Whereas some coping and adaptation measures are spontaneous and individually initiated, others are interventions from civil society organisations and national governments.

Like in agricultural communities, studies on gender and climate change in pastoral communities argue that adaptation differs by gender (IPCC 2007; Ngigi 2009; Djoudi and Brockhaus 2011). In other words, the differences in women's and men's roles, access to resources and status, result in differences not only in the vulnerability and impact but also in coping and adaptive capacity to climate change. Understanding the differences in copying and adaptive strategies to climate change by gender, and hence the influence of such strategies on livelihoods, is essential to chart ways on how to support pastoral communities cope better with and adapt to changes in the climate. This will significantly contribute to various MDG goals, including (and especially) MDG 1: eradication of extreme poverty and hunger; MDG 3: promotion of gender equality and empowerment of women; and MDG 7: ensuring of environmental sustainability.

Using Uganda, particularly Nakasongola District, as a case, this study examines the difference in coping and adaptive strategies to climate change of men and women in pastoral communities and how such strategies are influencing their livelihoods. Uganda's case is significant because of the gender terrain, and particularly as women in virtually all pastoral communities around the country are moremarginalised compared to their counterparts in agricultural communities.

Policies on socio-economic and political development in the country are more biased towards gender and women's concerns in agricultural communities than in pastoral communities. For instance, the current National Development Plan 2010/11-2014/15 (GoU 2010), which is the overall policy framework for development initiatives, is more silent on gender and women's concerns in pastoral communities than in agricultural communities. Even the Uganda National Gender Policy (MoGLSD 2007), is very explicit about gender and women's needs in agricultural community but implicit on pastoral communities. This means there is still not enough information about gender and livelihoods dynamics about people in the pastoral communities. Hence, in the context of climate change, it is vital to ascertain both impact and adaptation by gender. This would help in building strong institutions in pastoral communities that are sensitive to men's and women's needs. Specifically, therefore, the chapter looks at:

- The difference in the impact of climate change on men and women;
- The difference in the coping and adaptive strategies between men and women; and
- The influence of these strategies on the livelihood of the pastoral community.

Literature Review

In their paper on "Adaptation to Climate Change: Key Terms", Levina and Tirpak (2006) present and review definitions of various concepts of climate change such as *climate change, coping, climate variability*[6], *adaptation, and vulnerability*, among others. In the context of this paper only climate change, coping and adaptation are considered.

In their review, Levina and Tirpak (2006) present definitions of organisations, such as IPCC TAR (2001a), where *climate change* refers to any change in climate over time, whether due to natural variability or as a result of human activity; and IPCC TAR (2001b) where *climate change* is defined as a statistically significant variation in either the mean state of the climate or in its variability, persisting for an extended period (typically decades or longer). They further argue that *climate change* may be due to natural processes or external forces, or to persistent anthropogenic changes in the composition of the atmosphere or in land-use. Following Levina and Tirpak (*Ibid.*), UNFCC, Article I, defines *climate change* as a change of climate which is attributed directly or indirectly to human activity that alters the composition of the global atmosphere and which is, in addition to natural climate variability, observed over comparable time periods.

With regard to *adaptation*, Levina and Tirpak (*Ibid.*) note that the UNFCCC defines *adaptation* as the practical steps to protect countries and communities from the likely disruption and damage that will result from effects of *climate change*, while the IPCC defines *adaptation* as adjustment in natural or human systems in response to actual or expected climatic stimuli or their effects, which moderates harm or exploits beneficial opportunities (IPCC TAR 2001a).The Inter-governmental Panel on Climate Change (IPCC) lists the types of *adaptation*, including *anticipatory and reactive adaptation, private and public adaptation*, and *autonomous and planned adaptation*.

In their analysis of the definitions of *climate change*, Levina and Tirpak (2006) argue that while the first two definitions by IPCC (2001) take a broader view of *climate change* and the change is attributed to both natural variability and human activities, the UNFCCC definitions make a distinction between *climate change* that is attributable to human activities being responsible for the atmospheric composition of the global and climate variability attributable to natural causes. A common gap in the definitions is their disregard for gender, where all the definitions are stated in gender-neutral language. Subsequently, such gender-neutral language influences even how *adaptation* is defined as well.

Disregarding gender in the conceptualisation of such concepts masks an understanding of how gender dynamics are embedded in the vulnerability to the impact of, and thus, *adaptation* to *climate change*. A number of questions here come to mind. For example:

- Who among or between the men and women are more constrained by *climate change*?
- Given that the women provide over 70per cent of their labour to farming and they are already limited in access and control over resources, how is *climate change* impacting on them and what kind of coping mechanisms do they develop to ensure that their roles of food crop production are fulfilled?
- What is the effect of *climate change* on gender relations in the household and communities?

However, comparing the definitions of *adaptation* between the UNFCCC and the IPCC, Pielke (2004) notes that the UNFCCC definition of *adaptation* refers only to new actions in response to climatic changes that are attributed to greenhouse gas emission. Within the UNFCCC definitions, *adaptation* represents a cost of *climate change* and other benefits of these adaptive measures are not counted (Levina and Tirpak 2006).

On the other hand, from the broader IPCC (2001) perspective on *climate change*, *adaptation* policies also have benefits to the extent that they lead to greater resilience of communities and ecosystems to *climate change*, variability, and particular weather phenomena. Unlike the FCC, the IPCC (*Ibid.*) distinguishes several types of *adaptation*. In the interest of this paper, only *autonomous and planned adaptations* are considered because there is an interest in indigenous and institutional knowledge and strategies. However, as already noted, the lack of gender analysis of such concepts misses out an understanding of how *climate change* and variability differently affect men and women, and hence their adaptive capacity, and how this influences their livelihoods. Given their differences in gender, men and women are likely to perceive *climate change* and variability differently, including its impact and *adaptation*. It is in this regard that the subsequent section reviews literature on gender and *climate change* and *adaptation* with particular focus on pastoral communities.

Impacts of Climate Change on Men and Women

It has been argued that climate changes have different impacts on men and women due to their gender (Nampinga 2008; Djoudi and Brockhaus 2011). That is, the social relations between men and women with regard to power relations, division of labour, access to and control over resources and decision-making result in differences in the way climate change impacts on them. In the context of gender division of labour, for instance, studies indicate that besides women performing domestic duties such as cooking, child and elderly care, water and firewood collection, women in pastoral communities also take care of animals – performing tasks such as treating of animal diseases, watering animals and managing water sources (Omolo 2010; Secretariat of the Convention on Biological Diversity 2010). Men, on the other hand, are mostly involved in herding and marketing animals (Omolo 2010). Accordingly, in the event of prolonged dry seasons or drought, which results in water shortage, the labour burden women face is likely to increase as they will have to walk long distances to look for water not only for home use but also for watering animals (Olson, Rubin, and Wangui2010).

Men's role as herders implies that movements across their territorial boundaries in search of water and forage during low rainfall or droughts result in women being left to assume the entire responsibility of households and care for lactating cows and calves left behind for the period that men are not around. For instance, a study by the ACF International *et al.* (2010, 15) reported that among the Borana of Ethiopia, it is often the case that lactating cows and calves remain near the 'permanent' households, with the women and children, as the rest of the herd is taken in search of pasture by male members of the households. The same applies to the Turkana pastoral communities in northern Kenya (Omolo 2010).

The search for pastures and water often results in conflicts as one pastoral community ventures into or trespasses on another's territory (Nampinga 2008;ACF International *et al.* 2010; Olson, Rubin and Wangui 2010; Stark 2011). A study conducted in Mali and Ethiopia notes that droughts are leading to an increase in localised disputes due to struggle over access to water and pasture (ACF International *et al.* 2010). Findings in this study highlight that whereas previously conflicts among pastoral communities were about heroism and raiding of cattle, now they are about securing pasture and water (*Ibid.*).

Similarly, Oxfam (2008, 21) argues that resource competition among pastoralists is increasing risks of conflicts between different groups of land users. These risks are enhanced during times of stress (that is, during drought or floods). For instance, several clashes have been reported between the Turkana of northern Kenya and the Toposa of southern Sudan, when both are searching for grazing land and water in the Nadapal Belt (Omolo 2010). It is reported that in 2009, clashes between the two pastoral groups left more than 20 people dead and over 60,000 animals stolen within a period of three months (Daily Nation 2009b, cited by Omolo 2010).

In Uganda, climate change induced conflicts among and between pastoralists and agriculturalists have been reported in the cattle corridor districts of Luwero, Nakaseke and Nakasongola as well as in Karamoja region (Stark 2011). However, these studies do not assess gender-based conflict and how women in particular are affected by the conflicts.

Conflicts are not confined to only pastoral communities fighting each other and with agricultural farmers but also with other wildlife/ predators. A study in the region of Masai Mara, Kenya, reported a hyena density of 1.3 times higher on pastoral ranches than in the adjacent Maasai Mara National Reserve, whereas lion density was eight (8) times lower on pastoral ranches (Ogutu, Bhola, and Reid 2005). The higher density of hyena was linked to higher prey density in pastoral ranches while the lower lion density was partly attributed to conflicts with pastoralism (*Ibid.*).

Other conflicts with wildlife are noted to arise from the transmission of diseases from pastoral communities to wildlife (Secretariat of the Convention on Biological Diversity 2010). A study by Osofsky *et al.*(2005) reports that an outbreak of rinderpest in Kenya resulted in its transmission to wild grazers with death rates in the mid-1990s reaching 60per cent of buffalo and 90per cent of Kudu in some areas.

These conflicts have resulted in deaths of people, loss of animals and disruption of families within the affected pastoral communities (Omolo 2010). How this affects males and females, and their livelihoods, is not addressed by this study.

Deaths of people and loss of livestock resulting from such conflicts, partly accounts for the prevailing increase in poverty, deprivation, female-headed households (FHHs) and disruption of pastoral household livelihoods leading to social disintegration (Omolo 2010; Stark 2011). A study by KVRT (2009) in Turkana, northern Kenya, notes an increase in FHHs due to the death of a husband, separation/divorce, and mothers who remain 'single'. Consequently FHHs become more vulnerable to poverty than women in MHHs because of loss of or separation from a partner and also because women in FHHs cannot own livestock if they do not have a son nor can they employ a herder (*Ibid.*). Additionally, MHHs have also become vulnerable due conflicts which render their bread-winning role untenable.

In other areas, like northern Tanzania, it is reported that conflicts over water have led to a mass exodus of pastoralists, drastically reducing school attendance (Ngigi 2009).

In the context of ownership and control over livestock women usually own small stocks, with large stock being placed under men (Olson, Rubin, and Wangui 2010). However, even where women may have no ownership, they at least have control over milk and milk products (*Ibid.*), although such control may shift to men when milk production is commercialised (Kakwanzi-Kezaabu 2001; Kabirizi 2003). In the event of loss of cattle due to climatic changes, reduction in income and production is likely to ensue when the livestock, the main asset base of pastoralists, diminishes leading to food insecurity.

The impact of climate change on women, however, is acute compared to that of men. Unlike the men, women in pastoral communities are more vulnerable to the impact of climate change due to their subordinate position and their dependency on natural resources (Djoudi and Brockhaus2011). Virtually everywhere in SSA, women lack access to and control over most resources, including livestock and land. In the event of disruptive impacts of climate change on their livelihoods, women have low livelihood substitution possibilities compared to men (Maddison 2007), because men in most cases can seek casual work or semi-permanent employment. Even initiatives to empower and mainstream gender in pastoralism have continued to marginalise women due to their lack of ownership of livestock and limited opportunities for education and access to other productive resources (Ngigi 2009). A study by Oluka *et al.*(2005) in the eastern part of Uganda reports that the restocking programmes, as a result of cattle rustling, availed women an opportunity to enter the livestock production sector. However, as Oluka *et al.* (*Ibid.*) further note, women faced many challenges in the restocking process because livestock production is a culturally male-dominated enterprise. Although women provide the main source of labour for all the livestock production activities, such as tethering, kraal-cleaning and watering, they owned few animals. On another note, a study by Kabonesa (2000) in eastern Uganda reports that women's access to oxen has improved gender relations in households; for instance, women's access to communal land, women's participation in decision-making processes, particularly in family planning, and gender division of labour in favour of women.

Coping and Adaptive Strategies of Men and Women

According to Nyong, Adesina and Osman (2007), adaptation is when the individual or community cope with or adjust to the impacts of climate change.

The IPCC (2001a), on the other hand, defines adaptation as "an adjustment in natural or human systems in response to actual or expected climatic stimuli or their effects, which moderates harm or exploits beneficial opportunities".

Citing Parry *et al.* (2005), Ngigi (2009, 37) states that adaptation to climate change and variability entails actions that people take in response to, or in anticipation of, projected or actual changes in climate to reduce adverse impacts or take advantage of the opportunities presented by climate change.

Easterling *et al.*(2007) note two distinct forms of adaptation: 'autonomous adaptation' and 'planned adaptation'.

Autonomous adaptation refers to the ongoing implementation of existing knowledge and technology in response to the changes in climate experienced. This form of adaptation is reactive, considered as an initiative by private actors' response to climate change and variability (Malik, Xin and Smith 2010).

Planned adaptation, on the other hand, is the increase in adaptive capacity by mobilising institutions and policies to establish or strengthen conditions that are favourable to effective adaptation and investment in new technologies and infrastructure (Aguilar 2001; FAO 2003; Malik, Xin and Smith 2010).

According to a report by SACAU (2009, 29), "in climate change discourse, 'adaptation to' and 'coping with' climate change are terms that are often used loosely, and in some cases interchangeably". The report states that, "the Merriam-Webster Online Dictionary (2009) defines 'cope' as to, "deal with and attempt to overcome problems and difficulties" while 'adapt' is defined as, "to make fit (as for a specific or new situation) often by modification". Given these definitions, 'coping with' climate change would be more aligned to 'autonomous adaptation', as defined above, while 'adaptation' would relate more to 'planned adaptation". Despite the differences in the two forms of adaptation, it is worth noting that they are complementary, and some other authors have seen them as a continuum. Malik, Xin and Smith (2010) provide a more comprehensive analysis of autonomous and planned adaptations as well as other forms of analysis.

However, attempts to define and understand the meaning of adaptation to climate change have largely been confined to agricultural farmers and very rarely applied to pastoralists. Moreover, although many studies admit that gender influences adaptation, very few have comprehensively undertaken an analysis of adaptation from a gender perspective, let alone the concepts themselves, as noted above. Most studies have looked at women's vulnerability but are yet to pay more attention to adaptation.

Nevertheless, available studies on climate change and adaptation have observed that pastoralists have developed a highly flexible social system and an elaborate set of both individual and collective-based survival strategies (Omolo 2010). For instance, using findings from a study conducted by KVRT (2009) among the Turkana in the areas of Katilu, Kapua and Namoruputh, Kenya, Omolo (2010, 90) reports the following adaptive strategies among the Turkana pastoralist: increased mobility; herd splitting; redistribution of surplus livestock within social networks; diversification of livestock composed of camels, goats, sheep, cattle and donkeys to enable them to exploit different expanses of the range during any period of the year; livestock loans and gifts; formation of complex social security networks based on kinship and friendship;

reliance on relief food; fishing on Lake Turkana; farming along the rivers; and gathering wild fruits. Other coping strategies were noted to include: remittances from relatives; praying; joining merry-go-round groups by both women and men; shifting from pastoralism into agro-pastoralism, especially in areas situated next to water sources; seeking employment (mostly casual labour); and engaging in businesses, for example, selling firewood and charcoal as well as weaving baskets.

Similar coping and adaptive strategies have been reported among pastoral communities in Mali, Ethiopia, Uganda, Sudan and Somalia (Oxfam 2008; Olson, Rubin, and Wangui 2010; ACF International *et al.* 2010; Djoudi and Brockhaus 2011; Stark 2011). However, whereas some studies have been general in their analysis of coping and adaptive strategies others have endeavoured to integrate gender into their analysis although not comprehensively enough.

In the case of the Turkana cited above, although Omolo (2010) hardly offers a comprehensive analysis of such strategies from a gender perspective, she at least offers statistical information on the level of men's and women's involvement in every coping and adaptive strategy mentioned. For instance, in the Katilu areas, described as an agro-pastoralist area situated next to River Turkwell, where the key source of income was selling agricultural produce, the percentage level of women's and men's involvement was reported at 37 per cent and 42 per cent, respectively. In Kapua, where the main sources of income was selling livestock and livestock products, the percentage level of women and men's involvement was at 3.9 per cent and 21.9 per cent, respectively (*Ibid.*).

Agricultural farming was not an alternative livelihood in Kapua because the area was very dry, although another source of income was charcoal-burning by both women and men at 37.6 per cent and 41.1 per cent, respectively; while in Namoruputh, income from selling livestock products amongst women and men was 36 per cent and 48.8 per cent, respectively (*Ibid.*). In Namoruputh women were highly involved in selling of firewood at 20.4 per cent, while men were more involved in businesses at 14.6 per cent (*Ibid.*). The KVRT (2009) preliminary findings indicate that charcoal-burning is more prominent in dry areas with few or no alternative sustainable livelihoods. However, these findings do not indicate whether women control the proceeds from these strategies.

The percentage variations in women's and men's involvement in various coping and adaptive measures in the above study highlight the importance of gender in accounting for the differences in the coping and adaptive capacity between men and women in pastoral communities. For instance, where pastoralists embrace agricultural cultivation, the low level of women's involvement in selling of agricultural produce reflects much of the prevailing ideas that women in SSA hardly own and control land on which they cultivate their produce. Several studies argue that although over 80per cent of women engage in agriculture, implying that they have access either as wives, sisters, daughters, among others, such access is attained through men (FAO 2010; Rugadya 2010). Land is generally owned and controlled by men. Hence, even in pastoral communities, women usually have access to land but not ownership or control over it. But whereas women in MHHs can easily access land for cultivation as a coping and adaptive measure, women FHHs will face difficulties in adapting due to lack of access to land.

Embracing agriculture as a coping and an adaptive measure results in nomadic pastoralists settling in one place – what, in this case, we may conveniently call pastoral sedentarisation. By pastoral sedentarisation we mean a situation where pastoralists take up agricultural cultivation and confine herding of livestock in the settled territory. Pastoral sedentarisation, however, is likely to increase women's workload as the herds are kept close to homesteads and under women's care. According to Wangui (2003) as pastoralists diversify their livelihood to include cultivation, boys and young men become unavailable to herd livestock on a regular basis as they get enrolled in formal schooling. This information, however, is yet to be incorporated in the livelihood pastoral development policies, an oversight that, according to Hodgson (1999, cited in Wangui 2008), has produced unintended negative consequences.

The higher percentage of women involvement in charcoal-burning reflects how lack of access to resources and decision-making limits women's adaptive capacity, and instead makes them resort to measures that lead to greater pressure on natural resources, ultimately degrading the environment more. However, in cases where the sale or slaughter of livestock is the only option of coping in order to avoid loss or value of cattle through death and emaciation, this measure locks pastoral households in a pattern of depleting assets, thus challenging men's status, masculinity and power associated with having plenty of livestock.

Migration elsewhere in search of alternative work, although riddled with uncertainties, on the positive note, may open up new opportunities that make households more resilient to the negative impacts of climatic changes and may help empower marginalised groups such as women (Wangui 2008). For women, however, migration may be more uncertain since the limited access to resources and education mean they lack capital or the necessary skills for employment, thereby forcing some into prostitution. According to Hendrickson, Armon and Mearn (1998, 195), women and children are the first to leave the pastoral sector in times of crisis. They are sent off to stay with distant relatives or, ever more, to urban areas as street beggars where their vulnerability to food insecurity may not be relieved. In some cases, women who have moved to urban areas have been forced to turn to prostitution to survive (Omolo 2010).

Herders dispossessed of livestock are themselves often forced out of the pastoral sector into relief camps or into a search for wage labour (Hendrickson, Armon and Mearns1998). These gradual changes threaten the hopes of recovery as the crucial social ties needed to resume herding are often irrevocably severed (*Ibid.*). According to Barrow and Mogaka (2007, 14) such coping and adaptive strategies are increasingly transforming the socio-cultural and socio-economic organisation of pastoral societies. As men increasingly lose cattle, women have to play an active role to ensure family survival through engaging in diversified income-generating activities (Omolo 2010).

Whereas some of the above coping and adaptive strategies can be grouped under autonomous adaptation, planned adaptation, especially as interventions by civil society organisations and governments, have provided other options for pastoral communities to cope and adapt. For instance, there is an increasing campaign by various civil society organisations encouraging pastoralists to take on improved breeds into pastoral herds. The argument is that improved breeds have higher meat and milk yields, although they tend to be less drought and disease resistant, and sometimes need to be stall-fed (Olson, Rubin and Wangui 2010, 7). Stall-fed animals are particularly labour intensive for women as they are the ones to collect fodder and fetch water for the livestock as well as cleaning the kraals (*Ibid.*). At the same time, women may find difficulties adopting improved breeds given that culturally, livestock ownership in pastoral communities is largely patriarchal. Studies in the many pastoral communities have given evidence of women's limited access and ownership of livestock including lacking the power to make decisions about their own lives (*Ibid*; Djoudiand and Brockhaus 2011). As such, as Olsen, Rubin and Wangui (*Ibid.*) argue, for women in pastoral communities to acquire and access livestock requires careful negotiation and a deliberate attempt to meet with women in their spaces and at the convenience of the patriarchal structures within which the women live.

In some other contexts, adopting improved breed would mean selling many of the local breed to buy the improved breed. But as noted, since livestock ownership and control is under men and women lack access to resources, they may not have the money to buy the improved breed. A study by Chawatama, Mutisi and Mupawaenda (2005) notes that Zambian women cannot buy livestock because income from both livestock and crops is controlled by men. As such, enabling women to access enough cash income is instrumental in empowering them to cope with and adapt to changes and variability in climate. Where women have been able to take on livestock, the breed is usually small local stock, especially goats, sheep and poultry, that is well suited to local conditions and under their financial means (*Ibid.*).

In the context of shortage of water, several studies have proposed the need to adopt dams as storage facilities for use during dry spells (Ngigi 2009; Stark 2011). However, this has challenges related with competition from different pastoralists within the same locality, and may deepen conflicts. Besides, maintenance of dams may require technical and financial resources which may not be readily available in pastoral communities given that many of them are located in remote areas. Since men in pastoral communities control water sources, women may meet difficulties accessing such sources.

Whereas the above coping and adaptive strategies reflect men's and women's response within the socio-economic context, another crucial dimension is ecosystem adaptation. According to Vignola *et al.* (2009) ecosystem adaptation implies adaptation policies and measures that take into account the role of ecosystem services in reducing the vulnerability of society to climate change in a multi-sectoral and multi-scale approach. Thus, since pastoralism is a livelihood system tied to the ecosystem, it is vital to tap the local indigenous knowledge of both men and women in the preservation of the ecosystem. According to the Secretariat of the Convention on Biological Diversity (2010) women hold vital local knowledge regarding bio-diversity and make many decisions concerning the use of natural resources within pastoral households. Thus, mainstreaming gender into pastoralism-related policies, programmes and projects can contribute positively to bio-diversity conservation and adaptation.

Methodology

Study Area

This study was conducted in Nakasongola District, located in the north-western part of central Uganda. The district falls within the cattle corridor, and is described as one of the driest districts in the country with persistent droughts (Oloya*et al.* 2007; Government of Uganda 2009). The district is reported to be experiencing climatic changes and variability due to deforestation for charcoal-burning; re-allocation of land for agriculture; high stocking rate of livestock which is exhausting the pasture, grassland and other vegetation, thus land degradation; severe temperatures causing drying up of vegetation and crops; and bush fires.

Over 60 per cent of the population in Nakasongola rear livestock for their livelihood (Government of Uganda 2009), besides arable farming, retail and wholesale trade, charcoal-burning, fishing, brick-laying, local poultry keeping, bee-keeping, small-scale agro-processing and local construction work. The district is composed of eight sub-counties, including Lwampanga, Kakooge, Kalungi, Nakitoma, Nabiswera, Wabinyonyi and Nakasongola Town Council (*Ibid.*). Although there are many ethnic groups living in the district, the main ones are five. These are: Baganda, Banyoro, Baruli, Banyakore and Banyarwanda that are variously settled in the five sub-counties.

With the advice of district officials, Nabiswera Sub-County[7] was selected for data collection because it is one of the two sub-counties (besides Nakitoma Sub-County) where a majority of the inhabitants depend on livestock for livelihood. According to the 2002 population census, Nabiswera Sub-County had a population of 14,447, with the males estimated at 7,379 and the females at 7,068 (UBOS 2006). There were 2,831 households in the sub-county (UBOS 2006). In this area, livestock, wildlife and communities often share water from valley dams, valley tanks, seasonal ponds and streams.

According to the Nabiswera Sub-County Five Year Development Plan 2010/11 - 2014/15, the sub-county experiences rainy seasons from March-May and September -November. The rains are unreliable, though. The vegetation is typical savannah woodland with short scattered trees and grass. Seasonal streams criss-cross the sub-county. The area is generally flat with a few scattered hills.

Study Design, Sampling, Data Collection and Analysis

A cross-sectional study design was adopted for this study. It was considered appropriate given the limitation in the time and population coverage. Cross-sectional designs are suited, among others, for studies affected by such limitations. They are useful for conducting impact-related studies, and ascertaining an overall picture of the phenomenon as it stands – characteristics which define this study. And the design is cheap to undertake and easy to analyse (Kumar 1999).

Qualitative primary and secondary data were used for this study. Primary data were gathered through a rapid appraisal technique using focus group discussions (FGDs). To select the households from which data were gathered, purposive sampling was used. The process began with the identification of households that were engaged in pastoralism with the help of the Local Council Chairperson in the sub-county, followed by the random selection of individuals from those households to participate in the focus group discussions.

Data were gathered in two phases: the first data were collected in 2011 where two FGDs (one for women and the other for men) were conducted. Women FGDs constituted eleven participants while that of men was made up of twelve individuals. The second phase of data collection occurred in March 2012 wherein one FGD comprising of five men and six women was conducted. Primary data were also collected using key informant interviews held with three individuals who were purposively selected at the district and sub-county levels. Two of these included the District Veterinary Officer and the District Forestry Officer, and the third was the sub-county Community Development Officer (CDO). These were selected because their offices directly dealt with the community and environment-related issues in their respective fields.

In both FGDs and key informant interviews, the focus included the kind of climatic changes the community had experienced in the previous five to ten years; reasons for such changes; effects of such changes on men's and women's roles and their coping and adaptive strategies/measures; how such measures/strategies are changing people's livelihoods; organisations in the areas helping the community cope with the changes; and mitigation measures, among others. There were no civil society organisations/non-governmental organisations that dealt with pastoralists in the district.

Secondary data were gathered from government ministries, district, sub-county level, and NGO reports, libraries as well as internet searches on pastoralism and climate change in sub-Saharan Africa.

The study adopted a descriptive, explanatory and interpretative approach in the overall analysis of both primary and secondary data. The study was descriptive because it highlights men and women's experiences with climate change and the situation as it is on the ground. It was also explanatory and interpretive in that it assessed the implications of climate change on men and women's adaptive livelihood strategies.

Findings and Discussion

Impact of Climate Change

Several studies in different parts of Africa have admitted the reality of climate change and variability threatening people's livelihood and food security. Similarly, responses from participants in this study have pointed towards the same reality. All participants, in the FGDs and key informants acknowledged that Nabiswera Sub-County and the district in general were increasingly facing a lot of climatic changes and variability including long spells of dry seasons, with no rains for long periods and severe droughts in some seasons. Such changes account for the prevailing scarcity of pasture and water for livestock in the different parts of the district, and Nabiswera Sub-County in particular. According to the Sub-County Development Plan, 2010/11 - 2014/15, between 4-10 per cent of cattle die annually, depending on the severity of the droughts and/or dry seasons. This is in contrast to what some individual pastoralists have report. For instance, one of the participants in the men's FGDs stated that, "If one had over 100 cows, in the dry season one may remain with only 50". This was reported to take place mostly during the long dry spells. It was further noted that in most instances once the pasture and water are inadequate, instead of letting their animals die, some pastoralists opt to sell off most of the animals, and remain with a few which they can handle in such situation.

Underlining the severity and frequency of droughts, the District Veterinary Officer noted:

> Nowadays rainfall is erratic. It is unpredictable when it will rain. We used to have two rainy seasons each year but now it has been reduced to one. Pastoralists are now compelled to a nomadic lifestyle in search for pasture, water and food. But as they move around in search for pasture and water, conflicts tend to arise with our neighbouring districts, who are also suffering from the same agony.

Writing about climate change in Nakasongola, and Nabiswera Sub-County, in particular, Miti(2011) quotes the district Veterinary Officer to have stated:

> Due to mounting pressure on vegetation, pastures have reduced beyond 50 per cent. ...the rapid reduction of pasture is causing stress to animals and lack of pasture is currently responsible for about 30 per cent animal deaths in some places, and milk production has dropped by 60 per cent in some areas...if an animal happens to die at night, it is very possible to find it moulded the following morning. If there is no immediate intervention people will starve.

In light of the above, Miti (*Ibid.*) notes that on the ground, it is estimated that about 10 km^2 in Nabiswera have become bare. This situation heralds desert-like conditions if nothing is done.

The other impacts of the climate change were the spread of new diseases and plant species. As one male participant in the FGD stated:"Previously livestock such as goats were rarely treated or sprayed but now that has changed". Emphasising the increase in diseases, the participant further stated:"...as you may be immunising the animal against one disease, you are told to immunise against another outbreak of a new disease." Given that immunisation involves costs, many pastoralists were finding it difficult to meet costs related to the treatment of the animals.

It was also reported that droughts and long dry spells have increased the population of termites which are eating up the vegetation and degrading the land topography. Termites have affected more of the agricultural farmers because they eat up the cereal crops, especially maize when they flower and often attack other crops before they germinate. But even pastoral communities were affected since all the green pasture is also devoured by the insects. A District Forestry Officer noted: "The termites feed on anything they come cross such as grass and trees. The termites are contributing to low crop yields, loss of pastures, woodland degradation, soil erosion and silting of rivers and valleys".

Similar findings are reported by Stark (2011) in the Cattle Corridor, which included Nakasongola where testimonies from pastoralists indicated new invasive plant species, termites and other pests as well as increases in cattle diseases. In the same vein the sub-county CDO noted:

> Termites are more aggressive and destructive during the dry seasons as they search for moisture. These termites are greatly responsible for the dwindling pastures in the area.

The causes of climate change and variability in the district, including Nabiswera Sub-County, in particular, are attributed to human activities such as cutting trees/deforestation for timber, charcoal-burning or re-allocation of land for agriculture, overgrazing, and bushfires, among others. All participants in the FGDs indicated that the major reason why people engage in such destructive environmental activities is poverty, and the prevailing changes in the climatic conditions. Climatic changes have made rains unpredictable, hence, affecting agricultural productivity. With the scarcity of pastures and water, the cattle too become less productive. However, resorting to activities that are destructive to the environment leads to a cyclic pattern of problems. In other words, such human activities induce changes in the climatic conditions and, in turn, the changes in the climate force people to encroach more on the environment, hence, resulting in an environmental tragedy.

Furthermore, all the participants in the FGDs admitted that droughts were affecting their incomes and threatening their entire livelihoods as the population of their livestock dwindled. Although stated in general terms, the impacts of climate change, however, affect women and men differently in light of their gender division of labour, decision-making, and access to and control over production resources such as land and livestock, among others. For more on women's and men's activity profile, see Table 1.

Table 1: **Activity profile of men and women in Nabiswera Sub-County**

Activity	Men %	Women %
Digging	20	80
Grazing animals	60	40
Charcoal-burning	70	30
Fishing	90	10
Brick-laying	90	10
Looking for markets	85	15
Selling local brew	40	60

Source: Nabiswera Sub-County Five Year Development Plan 2010/11-2014/15

In other words, due to their differences in activity profile, men and women participants acknowledged that the intensity of the impact of climate change differed greatly by gender and also between female-headed households and male-headed households. For instance, one female participant stated:

> For us women, especially if one is a widow with children, you go through a lot to ensure that the children go to school and eat. We cannot afford to own many cattle. Even having a few affects us a lot; because besides looking for water and pasture, you also have to look for food. It is not easy being a woman nowadays.

Although the participants acknowledged the existence of gender division of labour, there was a shift of the burden of labour more towards the women. Like in other parts of the country, women in Nakasongola District have limited access to productive resources, such as land, livestock, information, credit and education. In the event of the crisis, women suffer more and their ability to extricate themselves out of such crises is met with a lot of obstacles.

In the above activity profile, although there are marked differences in the level of involvement between men and women in the different activities, women FGDs participants stated:

> In the event of cattle death or low milk production due to seasons without water and pasture, we women shoulder most of the food needs of the household members.

Traditionally, in virtually all pastoral communities, herding/grazing of animals was a responsibility of the men. But as climate changes, leading to death and low productivity of livestock, women and girls are forced to join in herding. This is because, as men in the FGD noted, climate change and variability is forcing many of them (men) to look for other sources of income. Female participants noted that the shift in herding to women and young girls happens in situation where a household's livestock asset base dwindles. In their own words, "Men see no need to spend their entire day looking after a few cattle and thus, they enlist the labour of young ones, both girls and boys including women".

During the journey to the sub-county, it was not uncommon to see young girls herding cattle, as can be noticed from Figure 1. Enlisting the labour of young ones in herding had implication on their schooling, and also risk for girls being raped. Even where children went to school, they had to graze the cattle after school.

Figure 1: **A young girl herding cattle, Nabiswera Sub-County**

Source: Photograph by the authors

To sum up, climate change and variability is increasing the labour burden of women and girls. Activities which were previously workloads of men are now shifting to women, in addition to their reproductive roles. The once common notion that men were the sole bread-winners is rapidly changing since, as can be noticed from the role profile, women too are actively involved in searching for household livelihood.

Djoudi and Brockhaus (2011, 128) report similar findings of increased women's workload among livestock and forest communities in northern Mali. They argue that several activities such as livestock herding and charcoal production, which were previously explicitly associated with men, were now shifting to women.

More than anything, however, such impacts tend to increase inequality and poverty among women and other poor men. The disastrous effects of climatic change and environmental degradation affect women and the poor most. This is because women and the poor tend to rely heavily on natural resources for their livelihoods; hence, their degradation negatively affects their welfare. For instance, as common property such as community forests where medicinal herbs, fuel-wood, berries and nuts, resin and gum are obtained, and green pastures, watersheds and rivers become scarce, this phenomenon perpetuates low productivity and entrenches women further into poverty (Rao 1999). According to Rao (*Ibid.*) there is a common inter-dependence between the negative effects of climate change and environmental degradation and women's welfare. As water resources become scarce, women get exposed to unsafe drinking water, poor hygiene and sanitation; as fuel wood becomes scarce, the time spent looking for it or for other alternatives causes under-nourishment which decreases women's potential for labour productivity, leading to high morbidity and inefficient use of their time (*Ibid*).

Coping and Adaptation

Adaptation to the impact of climate changes has been variously categorised. However, in this study, coping and adaptation is looked at from two perspectives: that is, autonomous adaptation and planned adaptation. As indicated in the literature review, autonomous adaptation refers to the ongoing implementation of existing knowledge and technology in response to the changes in climate experienced (Easterling *et al.* 2007). On the other hand, planned adaptation is the increase in adaptive capacity by mobilising institutions and policies to establish or strengthen conditions that are favourable to effective adaptation and investment in new technologies and infrastructure (FAO 2003). In most cases, planned adaptation, which normally is a response and interventions from civil society organisation, national governments and development partners tend to come after people spontaneously (that is, autonomous adaptation) either as individuals, households or groups find their own means of coping and adapting. Whereas some of such strategies may add more stress on the environment, others may be attempts to mitigate the escalation of the climate change and variability. And since adaptation is a function of living beings, with regard to humans, their coping and adaptive strategies differ depending on their gender, location or resources at their disposal.

Thus, in the context of autonomous adaptation, responses from FGDs and key informants at the district and sub-county levels indicated some of the following coping and adaptive strategies:

- sale or slaughter of cattle;
- diversification (by engaging in other income-generating activities such as charcoal-burning, business/trade, brick laying);
- supplement livestock with agriculture;
- sale of firewood;
- people digging their own ponds to help during the dry season; as well as
- temporary migration to river banks and lakes to graze and water the animals.

In what follows a more detailed explanation of these strategies is given.

Sale/Slaughter of Animals

This strategy has been adopted especially during long spells of dry season or drought. Instead of leaving the cattle to die due to the lack of water and pasture, men have opted to sell some of them. Decision to sell or slaughter animals has not always been consultative, especially in male-headed household with a woman as the wife or other female relatives. Participants in both the male and female FGDs said:

> Although in some instances men consulted their female counterparts on the decision to sell or slaughter the animals; in general, men often sold or slaughtered the animals without consulting the female counterparts.

The situation is, however, different for women in FHHs, who have the latitude to sell animals without consulting anybody. Participants in women FGDs, however, noted: "Although the herds were owned by the men, we at least have a say and control over the sale of milk and milk products and keep the proceeds".

The sale and slaughter of animals strategy is, however, short-term and has serious consequences for future livelihoods. In other words, as households sell or slaughter animals during drought or dry periods, this means that by the time they come out of such periods their herd has reduced drastically, hence threatening their livelihood and food security. Men whose masculinity is defined around possession of large herds of livestock, and the control of milk and products women exercise, dwindles and/or necessitates a shift to other activities. In other scenarios, the loss of livestock leads men to turn to idling, and drinking cheap brew or local gin, leaving women as the sole bread-winners. One female participant noted:

> Nowadays, it is difficult to understand the men. When life becomes hard, many of the men just abandon their families; they resort to drinking, idling about in the trading centre, and become more aggressive and violent at home. In the end, it is us, women who have to shoulder the burden of fending for the family members such as the children and older relatives.

Migration to River Banks, Lakes and Neighbouring Districts

This is another strategy adopted especially by pastoralists with large livestock. According to the Sub-County Development Plan, during drought or long spells of dry seasons, over 50per cent of pastoralists migrate to river banks and lakes such as River Lugogo, Lake Kyoga, River Kafu as well as neighbouring districts of Masindi and Luwero in search of water and pastures (Nabiswera Sub-County Five Year Development Plan 2010/12 - 2012/15). Migration to water bodies or neighbouring districts often leads to conflicts with other pastoralists and agriculturists especially when the former's livestock stray into gardens. For instance, one of the male participants in the mixed FGDs noted:

> Nowadays, herding cattle is becoming problematic. We are always in conflict with other pastoralists in the area, and our neighbouring districts as we search for pastures and water. Sometimes, if the cattle stray into some one's garden, it is even worse. If the crops are destroyed you are taken to the local authority, and you are told to compensate. And when the animals that have strayed to the gardens are captured, recovering them, you may pay a heavy penalty.

Participants noted that it was largely men and boys who migrate with the cattle, and women and girls are left behind. This is not only an attribute of gender division of labour but also reflects the fact of men's ownership and hence, the associated responsibility when it comes to large stock. Women are left behind at home to assume full responsibility of household needs. But this, as noted earlier above, increases women's workload and responsibilities, even though it enables women to become key decision-makers at the household, thereby enhancing their agency. On the contrary, men who are left behind do not necessarily become care-givers nor do they take on new domestic roles. A female FGD participant, for instance, stated:

> When pastures and water become scarce, men and young boys move long distance in search of water and pasture. They may take long to come back home. Sometimes they may spend over a week away from home. During this period it is us, women, who are left at home without any money for security. So we have to devise all means to ensure that we [get] food and other necessities for those left behind and ourselves.

Diversification

The other strategy involves pastoralists' diversifying by adopting agriculture farming as well as timber business, charcoal-burning and brick-laying. Participants in all FGD voiced that:

> As pastoralism becomes challenging many of us are resorting to supplementing with agriculture. Cultivating various crops such as maize, beans, soya beans, bananas, and pumpkins, among others, is becoming a necessity.

A woman participant added:

> When it comes to agriculture, it is normally us the women in agriculture, although men may help with clearing the bush and morning labour in the gardens.

Embracing agriculture for women depends much on women's access to land, which is largely influenced by relations with men. As noted in the literature review (FIG 2001; MoFPED 2006; Verma 2007), women have access to land but not ownership and control over it. Thus, pastoral households that have land may re-allocate part of their land for cultivation, although this depends also on the suitability of land for cultivation. In the context of gender, however, it is unclear whether the decision to adopt cultivation is consensual, that is, agreed upon between a woman and the man in the same household or not, and which part of the land to allocate for cultivation. Where household does not have enough land, it was noted that many hire or borrow it while others encroach on land around river banks and lakes or swamps.

Whereas in the context of hiring and borrowing, access results from a negotiated agreement between the land owner and one who is going to work on it, in case of encroachment on river banks and lakes access often results from seeking permission from the local authority who sometimes may demand some monetary rewards. In this regard, a female participant noted that many of them who do not have land have to rent land near lakes and rivers. But to secure access they are required to pay a certain fee, and depending on the amount of land, the price may range from 15,000 to 30,000 Uganda shillings. She added that, since many women find it difficult to access land due to lack of money to bribe or pay to cultivate land, some of them end up having small plots for cultivation.

Besides agricultural farming, pastoral households cope by resorting to cutting trees for timber, charcoal production and brick-laying. However, in the context of gender, men find it easier to engage in these activities, while women's reproductive and household chores mean that their involvement is on a limited scale. Cutting trees, charcoal production and brick-laying require a lot of time, search for markets as well as transportation. Such time is a luxury for many women, and the fact that generally women lack means of transport to take their items, for instance charcoal to trading centres, or lack social network to market their bricks, engaging their entire labour in such ventures becomes challenging for them. In most cases, where women are involved in such ventures, they will leave the marketing and transportation to be done by men. The following opinion from a female FGD participant is illustrative:

> For us women have a lot of activities at home. You have to care for the children and also look after animals. Although some of us engage in charcoal-burning, when it comes to selling or transporting the charcoal to trading centres, we enlist the labour of men and pay them something.

However, diversification involving any or all the above strategies does not necessarily mean that men and women's livelihoods are secure. In the dire need to cope, by adopting agriculture, for instance, agricultural communities too are being affected by climate changes and variability. Although even here informants noted that they try to adapt by using drought resistant seed varieties, they are having limited success. In the mixed FGD, for instance, it was noted that the district, through the sub-county, particularly the office of National Agricultural Advisory Services (NAADS) had provided them with seeds such as beans, maize, cabbages and tomatoes. But when rains take long, some of these items do not yield a lot.

A female participant in the same FGD observed:

> Some agricultural farmers especially those having banana plantation have devised a good way of watering their banana plantation. They get bottles of mineral water and fill them with water, and they put the bottles around each plantation, and insert holes in the bottles through which water drips to the plant.

According to Stark (2011, 35), limited success with agriculture despite use of seed varieties, "may or may not be the result of climate changes impacts, because those impacts also interact with severe and increasing environmental degradation resulting from massive deforestation for charcoal production, wetlands encroachment, and overgrazing. With the felling of trees for fuel wood, the balance of the natural habitat is disturbed, with unanticipated effects - for example, the birds that once fed on the trees now eat the farmers' crops."Additionally, though charcoal production and the cutting of trees for timber is a desperate need for men and women to secure survival, they are a reflection of poor natural resource management, hence, deepening the environmental problem of the community. According to Rao (1999), people in a desperate situation tend to be governed by their essential survival priorities. Their perception of risk management options tends to exacerbate any potential costs to themselves and to the ecosystem (*Ibid.*).

Engaging in Business

This strategy was mentioned by participants in all FGDs. However, the key question was what kind of business one could engage in, since among pastoral households, their business involves marketing cattle and cattle products such as hides and skins as well as milk and related milk products. Nevertheless, most participants said that some of them sell part of the livestock to start retail shops in the trading centres or engage in second-hand clothing business. Although this option is relatively easier for men to adopt, for women they lack the resources from which to derive capital to start similar ventures. Among married couples, although such businesses are started by the men, women too provide their labour, as in the sale in retail shops. In many cases, however, many women engage in petty trading selling items like tomatoes, oranges, pumpkins and other agricultural produce from their gardens. One male participant noted:

> In most cases, for some of us, men we have gone into business or casual labour. Women mostly engage in roadside businesses nearer to the homestead. These kinds of activities have helped a great deal.

Fishing

Fishing is another strategy among pastoral households as a form of diversification of their income. It is, however, done largely by men. Men are able to adopt fishing because culturally men's work activities are outside the homestead. It is an activity which requires people who are able to work away from home, in deep water lakes for long periods. Women with their reproductive and productive roles which require them to stay at home, find it difficult to fully engage in such activities. Where women are engaged in fishing, they are normally confined to selling the fish in the market, after purchasing them from those who capture them from the lakes and rivers, who are normally men. In the female FGD, one participant stated:

> As women, we cannot go fishing on the lake because that is regarded as a male activity. But even if we wanted, household work and looking after our children will be waiting for us to do. But for women who are involved in fishing, they normally buy fish from the men who get them from the lake, and then also sale in the local market at a retail price. Some women own boats, however, they hire them out to men who go fishing.

However, it was noted that fish in the lake, especially Lake Kyoga, is getting depleted as not only pastoralists were resorting to it, agricultural farmers and others too were going into fishing. Usually, when people resort to fishing, they do so without adequate knowledge of fishing, hence end up using bad methods. According to the Nakasongola District Environment Policy (Government of Uganda 2008) new fishing villages are emerging in the district. However, poor fishing methods are already threatening the sector, yet the Office of Environment charged with handling all environmental matters of the districts, is inadequately facilitated to ably enhance community participation in environmental resource management (Government of Uganda/Nakasongola District Council 2009).

Planned Adaptation

Whereas the above strategies were autonomous adaptation, the following are categorised as planned adaptation. They include, among others,

- adoption of improved animal breeds;
- tree planting;
- sensitising and training communities about the need to preserve the environment; and
- construction of valley dams.

Adoption of Improved Animal Breeds

Participants in all FGDs indicated that they are encouraged by both the district and sub-county authorities to adopt improved breeds of animals. In this regard, a participant from the male FGD indicated:

> Previously herding was free range, where we used to move from one grazing land to another without restrictions. But now with the increase in population demand for land and agriculture, a lot of people have gazetted their land. It has become difficult to graze in another person's land. Accordingly, we are being encouraged by the district and sub-county officials to adopt cross-breeds/improved breeds. Improved breeds are good because they produce a lot of milk; it is a better option instead of looking after a lot of cattle, with few benefits.

However, participants expressed some challenges with improved breeds. According to one female participant, "such breeds are labour intensive; they eat a lot, yet pastures are not available; and are easily vulnerable to diseases".

As noted in the literature (see for example, Wangui 2003), very often whenever pastoral communities adopt sedentary practices, the labour burden shifts more to women, as men move away in search of other income-generating activities. Women become charged with the responsibility of looking for pasture and water. Where pastoralists are sensitised to plant grass as pasture for the animals, the activity is often performed by women.

To facilitate the acquisition and maintenance of improved breed, study participants stated that programmes such as National Agricultural Advisory Services (NAADS) have given people top-up funds to purchase more cattle, especially the cross-breeds. They have also given them vaccines/drugs.

On the question of the selection criteria to give one funds to purchase cross-breed livestock, participants noted that the community through their various groups democratically select the beneficiary. Once the selected households get the improved breeds, the NAADS train and sensitise them about the care of the animal. According to female FGD participants:

> The sub-county started a dairy project, where one takes milk at the diary station and they record the litres one has brought, and at the end of the month, they are paid. This does not only apply to people with improved breeds but also to those with local breeds who may produce a lot of milk during the wet season when there is enough pasture and grass.

As already noted, in most cases, the burden of looking after such animals will always shift to women and children. Although most of the labour to care for such animal is provided by women and they may largely access and control milk and milk products, they do not own the animal itself.

Adopting improved breeds has some negative consequences; it means that pastoralists' local management and indigenous knowledge acquired over the years is discarded in favour of 'modern' management of improved breeds. Pastoralists' indigenous knowledge of ecosystem is lost since herding ceases to be migratory and instead confined to one area. As one female participant in the mixed FGD stated:

> Previously, people had developed a special way of understanding the local breeds, and treating them in case of illness. For instance, many of us who own local breeds we know each and every cow or bull by name. When you call one out of the kraal, the particular cow or bull will come towards you. This is unlike the improved breeds. We have to learn new ways of handling them. If there is something wrong with the local breeds, one will know by the unusual way it will behave.

In a related incidence, another female participant said:

> Over time most of our grandparents had developed knowledge of certain plant species that were medicinal and passed on the knowledge to us. But right now with the destruction of the forests all the knowledge is disappearing.

Similarly, Stark (2011, 36) argues that "in the context of reduced mobility, environmental degradation, and environmental change, indigenous knowledge has lost much of its force as a coping capacity".

Construction of Valley Dams

This strategy involves the sub-county through the district authority and with support from development partners helping the community construct valley dams as water storage for pastoralists. Participants in all the FGDs noted that because of shortage of water, the sub-county is helping the community by constructing valley dams that would help in lean times especially during long dry spells. Sometime, however, as participants further noted, often pastoralists are in conflict over the use of valley dams, not only with fellow pastoralists but also agricultural farmers since the dams are meant to serve the entire local community. The study by Stark (2011) in Nakasongola notes that pastoralists and farmers endeavour to make arrangements to accommodate their respective interests and avoid conflicts. The community form water management committees which provide opportunity to find constructive, peaceful solutions for the sharing of natural resources (*Ibid*).

Tree Planting

In periods of droughts, it is argued that pastoralists encroach on forests by cutting down trees for timber or charcoal production in a desperate move to secure survival. However, because this worsens the environment and the ecosystem, pastoralists and agriculturists are encouraged to plant trees. Here is what the Sub-County Community Development Officer had to say:

> The sub-county established fields for pine tree seedling; we purchase seedlings and distribute to communities and schools; and train the community in environment management and sustainability. Through Nalukonge Community Development Initiative Projects financed by the UNDP-SGF, the problem of termites and reducing of bare patches created by erosion is being tackled.

FGD participants have also acknowledged that the sub-county has sensitisation programmes designed to encourage people to plant trees by providing them with seedlings such as pine and eucalyptus. However, because these trees are also planted for commercial gains, women expressed some reservation, often stating that because they do not own land, it becomes a challenge for them to plant such trees. In that vein, one female FGD participant noted:

> The reason why sometimes we are reluctant to plant trees is because we are not sure whether one will last in the marriage. Trees take long to mature, by the time it matures the man would have chased you away [divorce], and get another one who is likely to benefit from your labour.

All the above forms of adaptations are indicative of how the impacts of the climate change in Nakasongola District, and Nabiswera Sub-County, in particular, is affecting their livelihoods. Nonetheless, which adaptation measure will lead to better result will depend on the extent of how women and men are able to mobilise and utilise resources within their reach. Whereas men are more likely to adapt better, the challenge remains for women. But not all women are affected and adapt in the same way. Depending on the resources at their disposal, women will adapt accordingly. Where women are unable to adapt, it is vital that government and other stakeholders intervene. Otherwise left on their own, such women and other vulnerable groups will be plunged further into poverty.

Conclusion and Recommendations

Like many other studies in other parts of SSA, the Nakasongola case proves the claim that gender influences men's and women's coping and adaptive strategies to climate change and variability.

Although the effects of climate change and variability are the same for women and men, how they are felt and the extent to which men and women respond differ due to differences in their roles, access to and control over resources at their disposal.

Unfortunately, despite the limitations in women's coping and adaptive capacity, impacts of climate change and variability add more burden by, for instance, shifting the once men's activities to women yet without any additional income. This situation, therefore, calls for urgent attention to women's plight as well as other marginalised groups who equally feel the pinch of climate change and variability.

However, male migration in search of alternative work, although increases women workload at the household in the short-term, in the long-term it gives women social space to assume autonomy and independence of household leadership, and allows them to get more involved in activities that were once strictly a male domain (Djoudi and Brockhaus 2011).

In general, however, there is a need for various stakeholders to urgently intervene to avert the impact of climate change, especially conflicts, reduce the work burden on women and girls, and also avert autonomous/spontaneous coping and adaptive strategies that exacerbate the environment such as cutting trees for the production of charcoal, timber and others that disrupt the ecosystem. Planned adaptation has to be sensitive to conditions of men and women, being conscious that even women and men are not homogenous. This implies, therefore, that response to climate change needs to be flexible, multiple and locally specific (Olson, Rubin and Wangui 2010).

Given that conflicts among pastoral communities go beyond the district boundaries and in other contexts go beyond the national boundaries, there is a need to hold consultative national and regional workshops to develop a national and regional environmental framework to resolve conflicts and mitigate environmental degradation.

Sensitisation of pastoral communities is significant to make them alert of the dangers of degrading the environment. Mere sensitisation, however, will not do much if pastoral communities are not provided with incentives or credit facilities that enable them to engage in environment-friendly income-generating activities.

This study also calls for the definitions and analysis of concepts for climate change, such as climate change, variability and adaptation, among others, to take into account gender considerations. Blanket definitions and analysis tend to perpetuate a situation where climate changes, its impacts and adaptation assume neutral repercussions for both men and women especially with regard to their livelihood options. Or at least such definitions and analysis of the concepts should be followed up with an explanatory note of the implications of the definitions on women and men's situation; because these concepts relate to men and women differently due to their gender.

Notes

1 **Climate Variability** - Climate variability refers to variations in the mean state and other statistics (such as standard deviations, the occurrence of extremes, and so on) of the climate on all temporal and spatial scales beyond that of individual weather events. Variability may be due to natural internal processes within the climate system (internal variability), or to variations in natural or anthropogenic external forces (external variability). See also climate change (IPCC TAR, 2001).

2 Nakasongola has five sub-counties, but Nabiswera and Nakitoma are the main sub- counties with pastoralists. In this case Nabiswera was selected.

References

ACF International, IDS, Tear Fund, IER, A-Z CONSULT, and ODES. 2010. Changing climates changing lives adaptation strategies of pastoral and agro-pastoral communities in Ethiopia and Mali. Institute of Development Studies (IDS), Action Against Hunger and Tear Fund Report, Brighton, Sussex, UK. Available from http://www.ids.ac.uk/idspublication/changing-climates-changing-lives-adaptation-strategies-of-pastoral-and-agro-pastoral-communities-in-ethiopia-and-mali.

Aguilar, Yvette. 2001. Autonomous and planned adaptation: The low watershed of the LempaRiver: El Salvador, Central America, UNFCCC-UNDP Workshop on adaptation methodologies. June 2001. Available from http://unfccc.int/files/meetings/workshops/other_meetings/application/pdf/elsalvya.pdf. (accessed on 26 July 2011)

Barrow, E. and Mogaka, H. 2007. Kenya's drylands – wastelands or an undervalued national economic resource. The World Conservation Union (or International Union for Conservation of Nature, IUCN). http://liveassets.iucn.getunik.net/downloads/kenya_dryland_value_2007.pdf (accessed on 14th July, 2011)

Chawatama, S., Mutisi C, Mupawaenda A.C. 2005. The socio-economic status of smallholder livestock production in Zimbabwe: a diagnostic study. *Livestock Research for Rural Development* 17(12). http://www.cipav.org.co/lrrd/lrrd17/12/chaw17143.htm (accessed on 11th July, 2011).

Daily Nation. 2009b. Never ending war between Turkana and Sudanese. In Omolo, Nancy A. 2010. Gender and climate change-induced conflict in pastoral communities: Case study of Turkana in Northwestern Kenya., AJCR Vol. 10. No. 2, African Centre for the Constructive Resolution of Disputes (ACCORD), Umhlanga Rocks: South Africa.

Djoudi, H. and Brockhaus, M. 2011. Is adaptation to climate change gender neutral? Lessons from communities dependent on livestock and forests in northern Mali. *International Forestry Review,* Vol. 13(2):123 -134.

Easterling, W. et al. 2007. Food, fibre and forest products. In M. L. Parry et al. (eds). Climate change 2007: impacts, adaptation and vulnerability. Contribution of Working Group II to the Fourth assessment report of the Intergovernmental Panel on Climate Change. Cambridge: Cambridge University Press, 273-313.

Food and Agricultural Organization (FAO). 2003. Gender and sustainable development in drylands: an analysis of field experiences. Rome: FAO. ftp://ftp.fao.org/docrep/fao/005/j0086e/j0086e00.pdf. (accessed on 12th July, 2011).

——.2010. Documenting the gender gap in agriculture, women in agriculture: Closing the gender gap for development.

Government of Uganda (GoU). 2010. National Development Plan (2010/11-2014/15), Ministry of Finance, Planning and Economic Development, Kampala, April 2010.

Government of Uganda/Nakasongola District Council and the National Environment Management Authority.2008. Nakasongola district environmental policy. Draft 03: May 2008, Nakasongola, Uganda.

Government of Uganda/Nakasongola District Council. 2009. Nakasongola District Environmental Policy, November 2009.

Hendrickson, D., J. Armon and R. Mearns. 1998. Vulnerability: The case of livestock raiding in Turkana District, Kenya. *Disasters*, 22 (3), pp. 186–195.

Hodgson, D. L. 1999. Images and interventions: the problems of pastoralist development. In Anderson, D.M. and V. Broch-Due, *The poor are not us: Poverty and pastoralism in Eastern Africa*. Athens, Ohio: Ohio University Press. Available from: http://www.cbd.int/development/doc/cbd-good-practice-guide-pastoralism-booklet-web-en.pdf. (accessed on 27th September, 2012).

Intergovernmental Panel on Climate Change (IPCC TAR). 2001a. Climate change 2001: Impacts, adaptation and vulnerability. IPCC Third Assessment Report. Cambridge: Cambridge University Press.

——.2001b. Climate change 2001: The scientific basis. IPCC. IPCC Third Assessment Report. Cambridge: Cambridge University Press.

——.2001. IPCC 3rd assessment report - Climate change 2001: Working Group II: Impacts, adaptation and vulnerability. http://www.ipcc.ch/ipccreports/index.htm (accessed on 15th July, 2011).

——.2007. Climate change 2007 – The IPPC fourth assessment report (AR4) http://www.ipcc.ch/pdf/assessment-report/ar4/syr/ar4_syr_spm.pdf (accessed on 15th July, 2011).

Kabirizi J.M., 2003. Improving food and feed security in intensive smallholder dairy farming systems in Masaka district. Proceedings of the 6th African Crop Science Society Conference held in Nairobi, Kenya, 12th - 17th October 2003

Kabonesa, C. 2000. Evaluation of post Beijing portfolio: DFID funding for gender in Uganda, 1996 – 2000.DFID Report

Kakwanzi-Kezaabu, R. 2001.Commercialization of milk production in households: a gender perspective. *Occasional Paper Series* 12. Department of Women and Gender Studies. Faculty of Social Science; Makerere University, Uganda.

Kankya, C, Muwonge, A, Olet, S, Munyeme, Demelash, M.B, Opuda-Asibo, J, Skjerve, E, and Oloya, J. 2010. Factors associated with pastoral community knowledge and occurrence of mycobacterial infections in human-animal interface areas of Nakasongola and Mubende districts,

Uganda. *BMC Public Health* 2010, 10:471. Available from <http://www.biomedcentral.com/1471-2458/10/471>(Accessed on 15th July, 2011)

Kenya Vulnerability Research Team (KVRT). 2009. Enhancing adaptive capacity of pastoralists to climate change induced vulnerability in northern Kenya. Preliminary findings of study undertaken between January 2008 and December 2009. Turkana, Kenya.

Kumar, R. 1999. *Research methodology: A step-by-step guide for beginners.* (1st ed.). London: SAGE Publications Ltd.

Levina, E. and Tirpak, D. 2006. Adaptation to climate change: Key terms, OECD, Paris. Available from: http://www.oecd.org/environment/climatechange/36736773.pdf. (accessed on 10th March, 2012).

Maddison, David. 2007. The Perception of and adaptation to climate change in Africa, *Policy Research Working Paper* 4308, World Bank.

Malik, A, Xin Q in and Smith, S. C. 2010. Autonomous adaptation to climate change: A literature review. *IIEP Working Paper* 2010-24. 20 August 2010.

Ministry of Finance and Economic Development (MoFED). 2006. Gender inequality in Uganda: The status, causes and effects, Discussion Paper II, August 2006.

Miti, J. 2011. Nakasongola turns bare as nature becomes foe, *Environment Journal,* cited in the *Daily Monitor*, 29 November 2011. Kampala, Uganda. (accessed hard copy as a pull-out) (Available from www.monitor.co.ug)

Ministry of Gender, Labour and Social Development (MoGLSD). 2007. Uganda national gender policy 2007. MoGLSD.

Nabiswera Sub-County five year development plan, 2010/11-2014/15, Nakasongola District, Uganda.

Nampinga, R. 2008. Gender perspectives on climate change commission on the status of women, fifty-second session. New York, 25 February – 7 March 2008.

Ngigi, S.N. 2009. Climate change adaptation strategies: Water resources management options for smallholder farming systems in Sub-Saharan Africa. New York: The MDG Centre for East and Southern Africa, the Earth Institute at Columbia University.

Nyong, A., Adesina, F. and Osman Elasha, B. 2007. The value of indigenous knowledge in climate change mitigation and adaptation strategies in the African Sahel. In *Mitigation and adaptation strategies andglobal change*, 12, 787-797.

Ogutu, J., N. Bhola, and R. Reid. 2005. The effects of pastoralism and protection on the density and distribution of carnivores and their prey in the Mara ecosystem of Kenya. *Journal of Zoology,* 265 (3):281-293.

Omolo, Nancy A. 2010. Gender and climate change-induced conflict in pastoral communities: Case study of Turkana in north-western Kenya. Available from: http://www.ajol.info/index.php/ajcr/article/viewFile/63312/51195. (accessed on 17th July, 2011).

Oloya, J. *et al.* 2007. Characterization of mycobacteria isolated from slaughter cattle in pastoral regions of Uganda. *BMC Microbiology*, 7, 95.

Olson, J. M., Rubin, D., and Wangui, E. 2010. Gender, agriculture, and climate change: A regional analysis for USAID/East Africa. Report prepared for USAID by DevTech Systems, Inc. Accessed from: http://www.usaid.gov/our_work/cross-cutting_programs/wid/pubs/ Gender_Agriculture_Climate_Change_East_Africa.pdf. (accessed on 25th June, 2011).

Oluka, J., Owoyesigire, B., Esenu, B., and Ssewanyana, E. 2005. Small stock and women in livestock production in the Teso farming system region of Uganda. Small stock in development (2005), 151. Research Project funded by DFID, UK. Available from http://www.dfid.gov.uk/r4d/Output/64499/Default.aspx. (accessed on 15 August 2011)

Osofsky, S.A.*et al.*(eds). 2005. Conservation and development interventions at the wildlife/livestock interface: Implications for wildlife, livestock and human health. IUCN, Gland, Switzerland, and Cambridge, UK.

Oxfam. 2008. Survival of the fittest: Pastoralism and climate change in East Africa. Oxfam

Briefing Paper 116, 2008. Online publication; http://www.oxfam.org/sites/www.oxfam.org/files/bp116-pastoralism-climate-change-eafrica-0808.pdf. (accessed 15 August 2011).

Parikh, Jyoti. 2007. Gender and climate change framework for analysis, policy and action: Integrated Research and Action for Development. Available from: http://www.undp.org/content/dam/india/docs/gnder_cc.pdf. (accessed on 21st March, 2012).

Pielke, Jr. R.A. 2004. What is climate change? Issues in science and technology. Summer, 1-4. Available from: http://sciencepolicy.colorado.edu/admin/publication_files/resource-486-2004.09.pdf. (accessed 21st March, 2012).

Perry, C.J. 2003. Water pricing: some important definitions and assumptions. *Occasional Paper* No 59, SOAS Water Issues Study Group, School of Oriental and African Studies, UK: University of London. Available from: www.soas.ac.uk/research/our_research/projects/waterissues/papers/38402.pdf. (accessed on 20th July, 2011).

Rao, P.K. 1999. *Sustainable development.*Malden: Massachusetts Inc, Blackwell Publishers Ltd.

Rugadya, M. A. 2010. Women's land rights in Uganda, status of implementation of policy and law on women's land rights for ECA, ACGS. Addis Ababa, June 2010. Available from http://www.aresearchug.org/reports/commissioned-research/Status%20of%20Implementation%20of%20Womens%20Land%20Rights%20in%20Uganda,%202010.pdf. (accessed on 20th July, 2011)

Southern African Confederation of Agricultural Unions (SACAU). 2009. Climate change: Key issues for farmers in Southern Africa, opportunities and responses. *Discussion Paper* No. 3, September 2009.

Secretariat of the Convention on Biological Diversity. 2010. *Pastoralism, nature conservation and development: A good practice guide*. Montreal.

Skinner, E. 2011. Gender and climate change: Overview report, BRIDGE Cutting Edge Pack on Gender and Climate Change, Institute of Development Studies. Available from: Brightonhttp://www.bridge.ids.ac.uk/go/bridge-publications/cutting-edge-packs/gender-and-climate-change/gender-and-climate-change&id=59217&type=Document&langid. (accessed on 17th July, 2011).

Stark, J. 2011. Climate change and conflict in Uganda: The cattle corridor and Karamoja. USAID Office of Conflict Management and Mitigation, Discussion Paper No. 3., Office of Conflict Management and Mitigation, USAID, Uganda. Available from http://www.fess-global.org/Publications/Other/Climate_Change_and_Conflic_%20in_Uganda.pdf. (accessed on 26 July 2011)

The International Federation of Surveys (FIGs). 2001.Women's access to land.FIG guidelines. Principles for equitable gender inclusion in land administration: Available from: Background report and guidelines. FIG Publication, No. 24 http://www.fig.net/pub/figpub/pub24/figpub24.htm. (accessed on 18th July 2011).

Uganda Bureau of Statistics (UBOS). 2006. 2002 Uganda population and housing census. October 2006.

UNDP-UNEP-Environment Initiative. 2004. Climate change, the environment and human welfare: Lesson learned from the Lake Kyoga catchment area. Policy Brief. Available from: http://www.unpei.org/PDF/Uganda-Policy-brief-Climatechange.pdf. (accessed on 10th March, 2012).

United Nations Framework Convention on Climate Change (UNFCCC). 1992. United Nations Framework Convention on Climate Change. UNFCCC, FCCC/INFORMAL/84; GE.05-62220 (E) 200705, Bonn. Available from: http://unfccc.int/resource/docs/convkp/conveng.pdf. (accessed on 10 March 2012).

Verma, R. 2007. Without land you are nobody: Critical dimensions of women's access to land and relations in tenure in East Africa. IDRC Scoping Study for East Africa on Women's Access and Rights to Land and Gender Relations in Tenure. Kenya, Ethiopia, Rwanda, &Uganda, January 2007. Available from: http://landportal.info/sites/default/files/ea_scoping_study_final.pdf. (accessed on 27th June, 2011).

Vignola, R., Locatelli, B., Martinez, C., and Imbach, P. 2009. Ecosystem-based adaptation to climate change: what role for policy-makers, society and scientists? *Mitigation and adaptation strategies for global change*, 14, 691–696., Turriaba, Costa Rica. Available from http://research.fit.edu/sealevelriselibrary/documents/doc_mgr/470/Global_Ecosystem_Based_Adaptation_to_CC_-_Vignola_2009.pdf. (accessed on 26th July 2011)

Wangui, Elizabeth E. 2003. Links between gendered division of labour and land uses in Kajiado District, Kenya. LUCID Working Paper 23. International Livestock Research Institute. Nairobi, Kenya. Available from: http://www.lucideastafrica.org/publications/Wangui_LUCID_WP23.pdf. (accessed on 15th August, 2011).

Wangui, E. E. 2008. Development interventions, changing livelihoods, and the making of female Maasai pastoralists. *Agriculture and Human Values* 25(3):365-378.Available from: http://link.springer.com/article/10.1007/s10460-007-9111-z?null. (accessed on 28th July, 2011).

Chapter 9

The Impacts of Climate Change and Variability on Pastoralist Women in Somali Region of Eastern Ethiopia

Bamlaku Tadesse[*]

Abstract

This paper focuses on the different survival or coping strategies developed by the Somali women to the harsh climatic conditions induced by the global climate change. In the study area, which is subject to periodic drought, men and women have different knowledge, management practices and exchange relationships. They have also developed complex adaptive strategies, differentiated by gender, to cope with seasonal climatic change, including spatial mobility, access arrangements to grazing and water resources, crop and herd diversification and management as well as exchanging relationships with sedentary and urban people. Climate change is likely to exacerbate both natural disasters and conflict over natural resources at all levels. Conflict situations are more severe on women than men. Women often become the sole care-providers for the young, the sick and the elderly. Crises force them into new roles (economic and social) that they may not be equipped for and are additional to their normal domestic burdens. Men and women often bring different skills to their households, communities and countries with respect to climate change adaptations and conflict management.

*bamlakutadesse@gmail.com

The international response to the implications of climate change has largely focused on mitigation initiatives (reduction of greenhouse gases), and has directed less attention to adaptation strategies, such as assistance with adapting to the adverse impact of climate change on food, livelihood, and human security. They have also displayed little regard for the social implications of climate change outcomes and the threats these pose for poor men and women, or for the ways in which people's political and economic environments influence their ability to respond to the challenges of climate change. To this end, the paper examines the problems caused by climatic change (that is, migration of youngsters, recurrent drought, food insecurity, degradation of their social capital, and so on) and their alternative strategies devised to cope with these diverse problems. In its descriptive analysis, the research has also tried to assess the status and position of women as well as their role in decision-making activities in the effort towards devising adapting mechanisms to climate change both at household and community level.

Keywords: *Gender, climate change, impact of climate change, Somali region, Eastern Ethiopia, adaptive mechanisms for climate change*

Introduction

Background and Rationale

The Somali region is one of the regions in Ethiopia severely affected by climatic changes in the Horn of Africa. The main climate-related hazards affecting pastoral and agro-pastoral communities in Somali region are drought, conflict, locust infestation, bush encroachments, livestock and human diseases, extreme heat, and land degradation (Devereux 2006). The region is arid and semi-arid and has repeatedly experienced droughts. Drought impacts include decreased pasture availability, leading to pasture shortage, overgrazing, and land degradation; decreased water availability, leading to water shortages and travel over long distances by women in search of water; crop failure in agro-pastoral areas; food insecurity and malnutrition, mostly affecting children, pregnant women and the elderly; abnormal community mobility; increased school drop-out rate due to community migration in search of pasture and water; interruption of development activities; increased human diseases and death (*Ibid.*).

With regard to community observations of climate change, there has been a shortening of the rainy seasons during the last decade (*Gu/Diraa* and *Karan* rainy seasons). Rains tend to start later and end earlier than before. In addition, rain frequency, distribution and predictability seem to have decreased in the region. This has led to scanty or no pasture growth; increasingly affected irrigation; and the depletion of resources (Zelalem and Fekadu 2009).

It has also been associated with rising competition and conflicts over pasture and water resources. A case in point is the conflict between the Gurgura clan of Somali with the Nolle clan of Oromia region over Erer Gota River basin (end and upper users, respectively). Both groups have similar interests in the river, that is, they have been practising irrigation on this river since the imperial time, that is, since the 1920s (Zelalem and Fekadu 2009). Due to the decline of the volume of Erer Gota River on account of the impact of climate change, both users now compete for the river-based resource, which invariably leads to conflict.

Although the decision to irrigate is based on choice, this choice is nonetheless influenced by climate. The current government of Ethiopia has offered pastoralists this option and some of them have accepted and practised it. As a result, the government has built irrigation canals as well as reservoirs for the underground water. Therefore, climate change has its own impact on the scarcity of natural resources especially on access to irrigation water which leads to ethnic or clan conflicts. Where ethnic or clan conflicts arise, women and children are the primary victims. This is because they lack the resources (or are disempowered) and unable to protect or defend themselves against attacks.

The impact of global climate change, population pressure, resource degradation, and scarcity of natural resources have greatly exacerbated the degree of vulnerability of the pastoral and agro-pastoral communities. As a result of this, the pastoral and agro-pastoral communities have developed indigenous survival (coping/adaptation) mechanisms to climate change and its impacts.

However, there are diverse forms of political, social, economic and cultural factors that influence and aggravate pastoralists' vulnerability to climate change. Resources are dwindling and the competition for such scarce resources which pastoralists used as input for their survival coping mechanisms is becoming intense. Such competition for scarce resources has not only led to further deterioration and mismanagement of resources but also resulted in recurrent conflicts and tensions. This, in turn, compounds the problem of resources scarcity and affects the livelihoods of pastoralists and agro-pastoralists. Such conflicts due to resource competition deter efforts of mobilising and effectively using the indigenous coping mechanisms to climate change as well as the efforts made by governmental and non-governmental organisations in the interventions to curb the impacts of climate change.

Problem Statement

Change on climate threatens everyone especially poor people, but it does not threaten all people equally (Patterson 2009). Women are disproportionately affected by natural disasters as they experience higher rates of mortality, morbidity and post-disaster diminishment in their livelihoods. This pattern of disproportionate impact has echoed throughout the entire world, and our world is a place where gender, class and climate change intersect.

Climate change is exacerbating the problems and inequities that women already face. Women are generally obliged to bear the burden of cooking food, collecting fuel and water, as well as to adhere to the cultural mores that prevent them from being educated, owning land or presenting solutions to village leaders (*Ibid.*). They are persistently affected by climate change because they generally do not have secure and affordable access to and control over land, water, livestock and other important resources. Elderly women, disabled women, and widowed women often face the most acute challenges related to climate change whilst having fewer resources to compensate for and adjust to changes (*Ibid.*).

Moreover, climate variability and climate changes impact the poor in developing countries more severely than those living in more prosperous nations (USAID 2007). Due to climate change impacts, the natural resources base of the region, particularly of the forest resources, has severely deteriorated and decreased. This is because human-induced adverse impacts exerted on these resources have been accelerated by climate change-induced shocks and hazards such as prolonged drought. Over the last three to four decades, the intensity of natural forest destruction and soil degradation has tremendously increased over wider areas of the region.

Furthermore, livestock feeds have desperately dwindled both in quantities and qualities. Areas, which were earlier used to serve as pasturelands for livestock grazing and browsing of animals in farming communities, have currently significantly shrunk and become farmlands because of great pressure imposed from cultivation. Generally, during drought, as there is less access to pasture, animals become weak and more susceptible to different diseases as they are forced to concentrate around a few water points.

In light of the above discussion, it is crucial to identify in these areas the political, social, economic and cultural factors influencing and aggravating people's vulnerability to climate change; the role of customary and modern institutions in the use and management of common natural resources, and their indigenous adaptation mechanisms to climate change and variability. These are central to design appropriate policy measures for better management and utilisation of the climate change adaptation mechanisms, scarce resources and the reduction of conflicts and tensions among pastoral communities.

This research has thus been designed to assess the impact of climate change on pastoral women and their livelihood diversification mechanisms and how this, in turn, has affected their effort to ensure their food security. The findings highlight ways of reducing food insecurity and poverty among the most vulnerable pastoral and agro-pastoral groups, mainly women and children. It is hoped that its results will influence policy makers to design appropriate conflict and resource management policies and strategies, to strengthen and promote existing indigenous knowledge, and to introduce new knowledge and skills into climate change adaptations. In addition, the study has been designed to evaluate the effectiveness of the existing coping strategies and to suggest better ways or alternative approaches and strategies.

Objectives of the Study

The general objective of this study is to conduct an in-depth analysis of the impact of climate change on Somali women and the strategies they have adopted to cope with the challenges. Its specific objectives are to:

- assess the social dynamics in Somali culture due to the influence of climate change on the traditional means of pastoral livelihood;
- identify the indigenous coping/adaptation mechanisms to climate change and its impacts; and
- examine the roles of customary institutions in the use and management of communal natural resources and their response to climate change.

Literature Review and Theoretical Framework

Ethiopia is extremely vulnerable to the impacts of climate change due to social, economic and environmental factors. In particular, high levels of poverty, rapid population growth, high dependence on rain-fed agriculture, severe environmental degradation, chronic food insecurity and frequent natural drought cycles complicate the vulnerability of this country to climatic changes.

Climate change has brought about a noticable impact on Ethiopia's temperature and precipitation: average annual temperatures nationwide are expected to rise by 3.1°C by 2060, and 5.1°C by 2090 (Aklilu and Alebachew 2009).

In addition, rain will decrease from an annual average of 2.04 mm/day (1961 - 1990) to 1.97 mm/day (2070 - 2099), with a cumulative decline in rainfall by 25.5 mm/year (Cline 2007; Aklilu and Alebachew 2009).

Climate change will have a significant impact on the quality of life in most of the least-developed countries (LDCs). It is estimated that by 2020, between 75 and 250 million people will be exposed to an increase of water stress due to climate change in Africa (Diarra 2009). The East African region is among the most food insecure in the world, and where climate change has the potential to aggravate the problem (Orindi 2005). Pastoralists face a series of challenges that hinder their way of life and stifle their ability to adapt to changes in their external environment. The pastoralists who inhabit the drylands of sub-Saharan Africa are among those who are living with the effects of climate change.

Pastoralists have been managing climate variability for millennia. However, the unprecedented rate and scale of human-induced climate change is beginning to pose more problems (Oxfam International 2008). Women and men in rural areas are vulnerable to climate change since they are highly dependent on local natural resources for their livelihood. Those charged with the responsibility to secure water, food and fuel for cooking and heating face the greatest challenges. Coupled with unequal access to resources and decision-making processes, limited mobility places women in a situation where they become disproportionately affected by climate change. It is thus important to identify gender-sensitive strategies to respond to the environmental and humanitarian crises caused by climate change. It is crucial to note, however, that women are not only vulnerable to climate change, but they are also effective actors or agents of change in relation to both mitigation and adaptation (UN Women Watch n.d.).

Adaptive capacity to climate change is the ability to adjust to actual or expected climate stresses, or to cope with the consequences (Elizabeth 2008). There is an overlap between disaster risk reduction and climate change adaptation. Adaptation to climate change is a long term adjustment to changing average climate conditions while disaster risk reduction is the management of risks and geo-physical hazards (Ryan 2010). In this connection, pastoralists across the Horn of Africa have long developed strategies to minimise the risks to their livelihoods and cope with predictable shocks in these arid and semi-arid lands. They have developed a system to adapt to climate variability and uncertainty and to cope with risks and reduce vulnerability.

Adaptation strategies are based on three pillars of the pastoral livelihood system: natural resources, livestock and people (Chinogwenya and Hobsen 2009). Such adaptation to climate change includes multiple responses, such as changes in crop management practices (fore example, choice of fields, planting dates, planting densities, and crop varieties); livestock management practices (for instance feeding and animal health practices, transhumance timing and destinations); land use and land management (such as fallowing, tree planting or protection, irrigation and water harvesting, soil and water conservation measures, tillage practices, and soil fertility management); livelihood strategies (for example mixture of crops or livestock produced, combination of agricultural and non-arm activities and temporary or permanent migration) (Elizabeth 2008, 8).

Adaptation can greatly reduce vulnerability to climate change by making rural communities better able to adjust to climate change and variability, moderating potential damages, and helping communities cope with adverse consequences. A better understanding of pastoralists' perceptions of climate change, ongoing adaptation measures, and the decision-making process is important to inform policies aimed at promoting successful adaptation of the pastoral sector. Adaptation calls for active and committed involvement of multiple stakeholders, including policymakers, extension agents, NGOs, researchers, communities and pastoralists (IPCC 2001).

Furthermore, the IPCC (*Ibid.*) refers to adaptation practices as actual adjustments, or changes in decision about their settlement areas/environment, which might ultimately enhance resilience or reduce vulnerability to observed or expected changes in climate. Adaptation to climate risks can be viewed at three levels (Muthukumara 2008, 4).

- firstly, responses to current variability (which also reflect learning from past adaptation experiences);
- secondly, observed medium and long term trends in climate; and
- thirdly, anticipatory planning in response to model-based scenarios of long term climate change.

Adaptation measures can be further classified into proactive and reactive or ex-post (*Ibid.*). Examples of proactive measures include crop and livelihood diversification, seasonal climate forecasting, community-based disaster risk reduction, famine early warning systems, insurance, water storage, supplementary irrigation and so on.

On the other hand, reactive or ex-post measures include emergency response, disaster recovery and migration.

Recent reviews indicate that a "wait-and see" or reactive approach is often inefficient and could be particularly unsuccessful in addressing irreversible damages such as species extinction or unrecoverable ecosystem damages that usually results from climate change (*Ibid.*).

Chinogwenya and Hobsen (2009,3) provide a good overview of the variety of pastoral adaptation strategies. Central to these strategies is mobility and herd management. Mobility is the foundation of this adaptive capacity, which ensures that livestock make the best possible use of dryland bio-mass. Resilience to climate change in the drylands cannot be supported if the institutions that have long underpinned the pastoral production system are marginalised. Mobility is, therefore, a key element in the pastoral production system, and experience has shown that if mobility breaks down, livestock production declines and environmental degradation ensues (*Ibid.*).

Mobility is not only about moving herds to areas of good pasture, but also about managing natural resources to ensure the existence of healthy pasture which will support these herds in the longer term. Management of mobility and resources has long been guided by customary institutions. These have been a vital part of pastoralists' adaptive strategy that helps pastoralists take advantage of opportunities and cope with the consequences of climate variation. Herd management practices that seek to accumulate and diversify herd composition and species during years with good rainfall performance, and migration in search of scarce water and pasture resources during bad years, are key strategies that have enabled pastoralism to succeed (*Ibid.*). Herd accumulation improves prospects for recovery by minimising the risk of unsustainable herd depletion through sales or increased mortality during bad years.

Four areas are considered as critical building blocks in response to climate change:

- mitigation,
- adaptation,
- technology transfer, and
- financing (Resource Guide on Gender and Climate Change n.d.).

Mitigation and adaptation are linked to manifestations of climate change while technology and financing are linked to the means for achieving development goals.

Mitigation involves a process of curbing greenhouse gas emissions from human activities - for example emissions from fossil fuels as well as deforestation - with a view to stabilising greenhouse gas concentration at a safe level (*Ibid.*).

Adaptation entails a range of activities to reduce vulnerability and build resilience in key sectors, such as water, agriculture and human settlements (Denton 2002).

New and improved technologies and financing initiatives at all levels also need to receive attention as part of collective efforts to address climate change (*Ibid.*).

Mitigation and adaptation efforts should systematically and effectively address gender-specific impacts of climate change in the areas of food security, agriculture, bio-diversity, water, health, human rights as well as peace and security (Resource Guide on Gender and Climate Change n.d.).

Migration is one of the oldest coping strategies for dealing with environmental change. Lately, however, due to perceived increase in the intensity and scale of environmental changes, many people see environmentally-induced forced migration as a new type of phenomenon (Odedra 2008).There are great methodological challenges in combining climate and migration models. Because of the multi-causality of migration, calculating the risks of population for migration due to the impacts of climate change could be considered as a hasty generalization. Some argue that climate change will trigger environmental conflicts that, in turn, trigger migration. Some, on the other hand, argue that migration may in itself trigger environmental conflicts. The two conflict assemblages relating to the environment and migration can be categorised as: first, climate change impacts on the environment → conflict → migration, and second, climate change impacts on the environment → migration → conflict (*Ibid.*).

The two main schools of thought in environmental conflict research are the neo-Malthusians and the Cornucopians (Gleditsch 2003). The neo-Malthusians claim that population growth leads to over-consumption of resources, resource degradation, and finally scarcities resulting in violent competition. Myers (1993) operates with premises that are similar to the neo-Malthusian. A good example of enquiry for this theory is the case of the intra/inter-ethnic-based conflicts between the Nolle Oromo in Karamille District, eastern Hararghe zone of Oromia Regional State, and the Gurgura Somali in Erere District Shinlle zone of Somali Regional State. The cause of their conflict is competition for irrigation canals, locally known as *magetta*. In the area, there is also conflict between the Issa Somali clan and the Afar ethnic groups, and between the Hawiya Somali and the Ittu Oromo ethnic/clan groups. All of these and other conflicts are due to the scarcity of natural resources like pasture/grazing lands, water points, farmlands, irrigation canals, and so on, besides other socio-cultural and ethnic/political dimensions. These conflicts often affect women, older people and children more than other sections of the community.

The cornucopians, on the other hand, emphasise the role of technology, human ingenuity and cooperation in overcoming scarcity. Cooperation rather than conflict may be the response to environmental challenges. Some researchers, among them Gleditsch (2010), claim that it is abundance rather than scarcity that often leads to conflict because rebel groups draw funding from the exploitation of natural resources and/or it is a conflict about the control of valuable resources. Black (2001) seems closer to the position of the cornucopians. While Somalia, Rwanda and Haiti are used as examples by the neo-Malthusians, those who claim abundance, rather than scarcity, fuels conflict point to countries such as the Democratic Republic of the Congo, Sierra Leone, Nigeria, Angola, Algeria, and Colombia (Odedra 2008).

Methodology

In this section, an attempt is made to describe the general socio-economic, physical and political situations of the study area in brief. Furthermore, methods of sampling design, data collection, types of data and data analysis are also described.

The Study Site

The research was carried out in the Somali Regional State, eastern Ethiopia where the impact of climate change on women is more severe and women are highly burdened with triple roles (that is, productive, reproductive and community roles). The study area is one of the most affected by the impacts of climate change such that its rangeland is almost unable to sustain the traditional modes of pastoral livelihoods. The recurrent drought in the region is a common phenomenon in which women and children suffer the most.

According to the 1997 census, the population of the region was estimated at 3.5 million, of whom 1.9 million were males and 1.6 million females (Devereux 2006). The region is linguistically and religiously homogeneous, that is, Ethiopian Somalis account for 96 per cent of the total population in the region, and 98.7 per cent of the population are Muslims (FDRE 1998). The Somali region consists of 9 administrative zones, 44 administrative districts, and 67 urban settlements.

The region is largely desert with high temperatures and low precipitation. Given the dominance of pastoralism and an ongoing shift towards settled agriculture, dependence on rainfall is more obvious and stronger today than in the past. The region has suffered a series of recent droughts and famines. The 1999/2000 and 2004 episodes, described by some as the worst in recent memory, led to numerous deaths both of humans and livestock (Devereux 2006). It also led to widespread poverty and displacement of many pastoralist families. Although the drought and famine did not rise to the level of the "media famines" of the 1970s and 1980s, these rounds of drought and famine have also resulted in severe livelihood shocks. The region's economic fortunes are also intertwined with and are affected by local, regional, and even international changes in economic circumstances which, for this region, are always changing, and in ways that are difficult to predict (*Ibid.*).

Sampling Design and Method of Data Collection and Analysis

This study has been conducted in the Somali region of eastern Ethiopia on selected districts, namely: Afdem and Erer in Shinlle Zone, and Harshine and Awebare in Jijiga Zone. The study has employed mainly qualitative design in which data were collected using semi-structured key informant interviews, case studies, focus group discussions, observations and informal discussions. These methods of data collection have been used to collect qualitative data on human perceptions, attitudes, beliefs, and so on towards the impact of climate change on pastoralist women. Supplementary information has also been collected from 202 households using a household survey in selected districts and peasant associations (PAs). Two districts in each zone listed above and two PAs from each district were selected purposively based on the severity of the impact of climate change and the degree of women's livelihoods diversification in response to it. In-depth interviews with key informants were conducted and data generated in order to have a thick description of the indigenous coping strategies to climate change. Key informants were selected purposively based on their sex, age, level of education and awareness/knowledge about their culture, society and environment. Parts of the study population are evaluated in the context of the whole system, that is, a holistic approach. Purposive and snowball sampling techniques were employed to select the key informants from each PA. The number of key informants (a total of 16 consisting of 10 females and 6 males) was adjusted after factors and conditions became clear and directive while the study was in operation. In addition, a total of eight focus group discussions (two focus group discussions in each PA, one with men and another with women household heads) were conducted to generate data for the purpose of understanding the various indigenous coping strategies to climate change and related issues. The focus group discussion participants were selected purposively based on their sex, age, level of education and awareness about their environment. In addition to collecting primary data, there were also systematic examinations of relevant documents – both published and unpublished – as secondary data sources.

With regard to analysing qualitative data, the proposed research was based on the theoretical and methodological principles of subjective interpretations. To this effect, a content analysis technique was employed to analyse the qualitative data, and the quantitative data were analysed using descriptive statistics.

Results and Discussion

This section of the paper tries to illustrate the main theoretical and empirical discussions of the study. It focuses on some of the indigenous adaptation mechanisms of Somali pastoralist women to climate change and variability. The discussion also includes the impacts of climate change on social dynamics in their culture, and on the livelihoods of Somali women and resource-poor households.

Impact of Climate Change on Social Dynamics in Somali Culture

Elders in focus group discussions pointed out that climate change has caused customary institutions to lose their significance and modern institutions were not ready to take their place. This is especially relevant in regard to land grabbing for the establishment of private enclosures or for appropriation of communal land for charcoal-burning, where customary land-use agreement no longer applied. In this sense, climate change does not only lead to erosion of customary institutions, but also to an overall disorientation of environmental governance. This in turn leads to the rising level of conflicts, especially on land, and further environmental degradation, since degradation of communal land can mainly be seen as the consequence of the collapse of customary institutions.

Furthermore, institutions for rangeland management broke down with the introduction of modern institutions. Government policies on land tenure regarded rangelands as communal and open to all pastoralists. In the region, rotational grazing system on rangelands was highly practised. This is where parts of the pastoral lands were kept closed off to grazing and utilised only at certain times of the year to enable the recovery of the other lands which also reduced the impacts of droughts.

However, there is a considerable shift in the livelihood system from pastoralism to agro-pastoralism. This has created a competition between the two modes of production. It has further aggravated land conflicts, for example, among the Issa and the Ittu clans in Afdem and Meiso districts where the pastoral and agro-pastoral communities interface. This, in turn, has exacerbated the establishment of enclosures (private grazing and farm lands) which has played another negative role.

These customary systems demonstrate that dealing with hazards is not a new challenge among the members of the communities focused on by this study. However, with droughts becoming more and more severe and frequent, the redistribution systems are becoming impractical as the number of people needing social support is increasing from year to year. The effectiveness and sustainability of customary resource management systems are also threatened by increased drought frequency, due to abnormal mobility and exacerbated resources. Social associations, such as youth associations, inter-youth support groups, and women's income-generating groups, are also negatively affected by drought impacts. Increased migration and longer durations spent searching for water, livestock feed and food allow little time for social activities and meetings (Bamlaku, Yeneneh and Fekadu 2010a). Drought also impacts on savings and credit groups/cooperatives negatively by reducing incomes from livestock, livestock products, crops and other activities such as petty trading. Government officials in Erer District mentioned that there is an increasing rate of drop-out members from saving and credit cooperatives due to recurrent droughts. Despite being affected by drought, all of these social institutions remain important to people's adaptive capacity.

The Somali women play a crucial role in the socio-economic activities of society. The division of labour is clearly defined and heavily weighing the women down. Traditionally, a Somali pastoral woman milks animals, processes the milk, feeds the family, and cares for and watches the livestock (Bamlaku, Yeneneh and Fekadu 2010b). She also collects firewood, cooks, feeds the children, cleans the house and washes the clothes and the utensils. In addition, the women bear the responsibility of dismantling and rebuilding their *dessa* (traditional huts) as they move from place to place in search of pasture and water for their livestock. Figure one depicts that process.

Figure 1: **Women dismantling and reconstructing their *dessa* in Harshin District**

Source: Photos taken by the Author

Concerning its impact on gender division of labour, climate change and the resulting loss of livestock, water and vegetation, has changed the traditional workload. Currently, firewood and water collection takes much longer than previously. Similarly, the preparation of food can be more tedious since ingredients are getting scarce. Where livestock is lost, many women have to replace the care for animals by other domestic or income-generating activities like selling tea and *khat*, and other strategies. This helps women to be economically empowered and influential in decision-making processes at household and community level. The preparation of ropes from plastic bags (instead of acacia fibre) for household use is one that has been recently adopted; the other is the preparation of rooftop material for their *dessa*.

Furthermore, the declining number of livestock during droughts also affects the number of marriages, as some livestock is needed as a dowry, a practice that affects reproduction. Traditionally, marriages and festivities are conducted during the rainy season near rivers where there is abundant numbers of tree shades. If rivers run dry for longer periods due to the prolonged drought, such ceremonies will be postponed until the rain comes (Bamlaku, Yeneneh, and Fekadu 2010a).

Climate change, along with other social and environmental factors, have changed pastoral livelihoods in various ways. This has affected women's livelihoods, their coping strategies and their social position in a number of ways.

Firstly, it affects pastoral livelihoods by increasing the scale of poverty and the disruption of families, particularly among the poorest groups of pastoral families. This further aggravates the magnitude of poverty due to the decline of the natural resource base induced by climate change. This in turn leads to a disruption of families as men migrate to towns to look for jobs while women are left behind to take care of the children and the remaining livestock. Usually this situation is a desperate action for women and children because they have only a few animals left behind and also have to live in a small *dessa*.

Secondly, there is an increasing rate of pastoral mobility due to the recurrent drought and the decline of natural resources. In this case, women are left behind more frequently, although on a temporary basis. On the other hand, there is also a decrease in pastoral mobility as a response to the declining nature of the resource base that pastoralists heavily depend on due to acute shortage of water and animal diseases. This, in turn, forces them to settle in a permanent area which does not require much movement.

Thirdly, as a result of climate change, pastoralists' way of life is undergoing great alteration. The trend is moving towards higher vulnerability, loss of solidarity mechanisms for coping with droughts, destitution, and dropping out of pastoral life. Pastoral production system depends on the availability of natural resources that are sensitive to climate change. With the prevailing vulnerable situation of pastoralists as well as future uncertainties threatening over the whole spectrum of pastoral production, many households have opted for some sort of diversification in order to fill existing economic gaps caused by climate change. Such adaptation strategies include the selling of milk, engaging in petty trade, charcoal production, and out-migration to new areas for wage labour, among others. Nowadays, many women are organised in milk selling cooperatives. However, milk production and milk sales are negatively affected by the declining nature of the resource base due to the increasing rate of clearing the vegetation cover for charcoal trade which, in turn, adversely affects their income from milk sales. Selling milk in the past was uncommon to an extent that any person or family involved in selling milk was nicknamed as milk-seller. Traditionally, surplus milk was always kept and freely given to guests and travellers. Nowadays, selling milk is a well organised business with well established networks of collection by women. It has now become an important contributor to pastoral household income.

Besides, charcoal-burning has become a new strategy for coping with economic losses from climate change. Nowadays, the economic returns from charcoal production are competing with that of livestock production. Charcoal production has become a fallback option for poor pastoralists who are unable to carry on with their traditional way of life. The activity is widely practised in the study area. As discerned from discussions with community members, it remains the most important option for generating alternative income. This can be evinced from the number of people involved, coverage, market opportunity, the low investment required to kick start it, and the free access to tree resources. The widening and increasing consumption of *khat* (a mild stimulant leaf) among pastoral youth and the deteriorating livestock herd sizes make them resort to charcoal production. As a result of the recurrent drought conditions, bush encroachments and the consequent decline of pasture and water, among others, pastoralists' livelihood strategy is greatly jeopardized due to the declining nature of livestock herd sizes. This in turn aggravates the unemployment rate of pastoral youth since the accumulation of livestock is a source of employment for the pastoral household. These unemployed pastoral youth are engaged in charcoal production and selling to urban centres. Because of this network with urban centres and the consequent cultural diffusion, they are also accustomed to the habit of chewing *khat* as they will have extra-time to engage in it. However, almost all the income from charcoal will not be used for the benefits of the families; some of it is used for purchasing *khat*, the consumption of which helps men to overcome the misery and feeling of hopelessness they are exposed to due to the precarious future of pastoralism.

Furthermore, an increasing number of pastoralists who have lost their animals are leaving livestock production altogether, making their way to major urban centres, particularly to Dire Dawa, Harar, Jijiga, Djibuoti, and Hargeisa towns. The combination of increased climatic shocks, practices which hinder mobile pastoralism like the introduction of regional boundaries, and the lack of other viable livelihood options is pushing more and more pastoralists out of the system, many of them ending up in urban areas. Their remaining herds are left behind under the care of the women, elderly family members and children. In a way, this is some sort of coping strategy whereby family members who succeed in getting employment in the urban areas support their families in the rural areas.

Indigenous Adaptation Mechanisms to Climate Change among Somali Women

Regarding the contextual definitions and community observations of climate change, there has been increasingly a shortening of the rainy seasons during the last decade. That means rains usually start later and end earlier than before. In addition, rain frequency, its distribution and predictability seem to have decreased in the region. This is leading to scanty or no pasture growth, increased water scarcity-impacted irrigation, and depletion of resources. Due to the absence of rain for longer periods of time (which could be for a year and even longer), most important plant and grass species are becoming extinct since their seeds are no longer able to germinate when the rain comes. As a result of this, it is observed, the previously pastureland is now turned into barren land full of dust particles.

There is also an increasing rate of bush encroachments (unwanted trees and weeds which are not edible by animals) that is replacing the pastureland. According to the community elders, these encroachments are the result of climate change which has a devastating effect on the availability and rehabilitation cycle of pasturelands. In other words, pasture lands are gradually being replaced by these unwanted bushes and weeds, which is also leading to increased competition and conflicts over pasture and water resources.

Adaptive capacity is defined as the potential or ability of a system, region, or community to adapt to the effects or impacts of climate change (Smit and Pilifosova 2003). It is the ability or potential of a system to respond successfully to climate variability and change. It includes adjustments in both behaviour and in resources and technologies to reduce exposure to climatic hazards, and enhance the capacity of systems to cope with its impacts. Determinants of adaptive capacity include the availability and distribution of resources, available technology, structure and function of institutions, human capital, including education and personal security, social capital, including property rights, ability of decision makers to manage information, and the public's perceived attribution of the source of stress (IPCC 2001; 2007). These attributes will differ among regions, communities, and individuals and will vary over time, translating into different capacities to adapt (Adger 1999).

There has been a continuous drought for the past three years in the study area. Men migrate with animals to places where they find pasture and water. And those who lost all their livestock rely on relief aid from governmental organisations and NGOs. The relief is distributed through a committee constituted by community members. NGOs encourage women's participation in the committees. Other coping strategy is the migration to nearby cities to become wage labourers. The focus group discussion participants also indicated that there are no early warning systems for drought. The FGD participants maintain that community members mostly migrate and rely on better-off relatives. However, drought coping strategies, such as engagement in different off-farm activities or relief aid, are rarely noticeable in the community (FGD with selected elders from Bedhadale PA, Awebere district on 22nd July 2009).

Women play an important role in the pastoralist way of life. They hold diverse responsibilities with regard to livestock, agricultural land and the household since they develop an intimate knowledge of natural resource management. These enable them to put all this into practice for the benefit of both their communities and the environment. However, their knowledge and capabilities have not been fully recognised, and they are often excluded from the decision-making process at all levels. In times of scarcity of natural resources, while stress and hardship rise for everyone, it is the women who are most burdened with the increased workload as they struggle to compensate. Moreover, their ability to respond to economic opportunities is often constrained by traditional beliefs about gender roles in pastoral societies. In general, women lack time, financial resources and the networks necessary to take advantage of opportunities. Given the chance, they have proven to be highly capable of finding alternative sources of income, particularly during periods of intense drought in the drylands, and of acquiring the necessary skills to do so. As keepers of traditional knowledge, women pastoralists make an important contribution towards the sustainable management of land and natural resources (Ahmed *et al.* 2008).

Women carry out a multitude of tasks involving close interaction with and dependence on the natural environment. As a result, they have been the main custodians of the Somali indigenous knowledge systems by means of which they conserve and are sustained by their natural resource base. Since women are charged with the responsibility of rearing children, they also pass on this knowledge through songs, riddles and proverbs to the younger generation, thereby contributing towards future environmental conservation. As house building is women's responsibility, they do not cut down live trees, but cut only the branches or collect dead wood. The preservation of trees and plants is very important since they have a wide variety of uses. A Somali woman knows a variety of species of plants that can be used as medicine for both humans and livestock, for insecticides, for house building and for rituals. However, they are not allowed to participate in such ritual ceremonies together with men (Bamlaku, Yeneneh and Fekadu 2010a). The FGDs revealed that there is some indigenous knowledge to control livestock diseases among Somali pastoral communities. Most of these are from plants' leaves and roots. Women have high contribution in the preservation of livestock products that may stay longer ranging from a month to a year. One of the best examples identified at Aydora is *Satte* (preservation of food made of sorghum mixed with butter). To make the food stay unspoiled for longer period with this system, women use leaves and other parts of different plant species (FGD participants of elders from Aydora PA, Erer District, 20th July 2009). Certain trees and plants have special qualities in the preservation and flavouring of milk, which is the staple diet of pastoralists. Splinters of the wild olive tree are burnt and the charcoal is used to smoke the milk gourds, both to sterilise the milk and to give it the smoky olive flavour which the Somali enjoy. The wild olive tree is preserved by women for this purpose.

Climate change adaptation strategies will be necessary at all levels, from internationally funded projects and national government policy responses to community-level projects and household coping strategies. To the extent that these initiatives build resilience and response capacity, bolster human security, and reduce grievances, they can be instrumental in avoiding climate change-induced conflict. In government-level adaptation, policies implemented by governments can help strengthen community adaptation and significantly boost resilience. Investing in technological adaptations, such as water harvesting and irrigation works, sanitation systems, and early warning systems, may help developing countries better cope with climate change. Reliable water for irrigation can help blunt the effects of drought. A country-wide network of weather monitoring stations can give policy makers data about climate change that increases their capacity to make informed interventions. Measures for technological adaptation may be expensive and involve a high degree of coordination in order to be developed and maintained effectively (Jeffrey, Christine, and Kelly 2009).

At the community level, adaptation responses are likely to focus on ways to increase livelihood diversification and food security. Such strategies may include establishing seed banks and facilities for food storage/preservation that can provide livelihood support for pastoralists and provide a buffer in times of food shortages. Communities develop local level irrigation schemes that they can implement and maintain by diverting local water sources with the help of channels or small generators. Communities also adopt rain harvesting techniques that will allow them to collect water during downpours, thus allowing them to release water as may be required during dry periods.

Climate change is likely to exacerbate land degradation and intensify the prevalence of pests.

Climate change will also require adaptation measures at the household level. Individuals will have to adjust livelihood strategies that are no longer functioning as viable coping mechanisms or may even increase insecurity. Such adaptations include agro-pastoralists' ability to choose drought resistant indigenous crops. Dependence on livestock may no longer be a viable strategy in places where dry conditions and a lack of water and pasture are likely to worsen. Besides the commonly practised adaptation strategies, there is charcoal production, petty trading, utilising child labour, marriage alliances with better off families/clans, leasing out land, benefitting from relief grain, sending children to economically better families during food shortage, selling of large animals, borrowing grains, migrating to either urban or nearby rural areas for wage labour, and so on.

In cases where households are unable to adapt their livelihood strategies sufficiently to threats to their food security, many families are simply forced to reduce consumption. This may be manifested in reduced caloric intake. In this regard, the household survey shows the following results concerning the number of meals per day and variations in meals frequency per season.

Out of a total of 202 respondent households, only 4 per cent reported to have had one meal per day. While 56.4 per cent of the respondents reported having two meals a day, the remaining percentage (37.6 per cent) of the respondents mentioned eating three times a day during periods of food insecurity.

With regard to variation in meals frequency in each season, the majority (67.8 per cent) of the respondents replied that there were variations during the dry and wet seasons. The remaining (29.2 per cent) respondents were of the opinion that there was no variation in meals frequency per season.

Furthermore, there always exist social support mechanisms to community members who are in need of help. For instance, there is the credit (in cash) type of assistance for those who are able to work and return the credit. Moreover, FGD participants indicated that there is also credit service (in type) where grains may be borrowed from neighbours when they receive relief assistances.

There is no exclusion for getting social support on the basis of clan identity. Nonetheless those who violate the norms and values of the community may be excluded from the support system, that is, less assistance is extended to such members (for instance, assistance may only be forwarded only when they face a chronic problem). Such supports might be also extended in times of mourning, weddings, and drought. Besides, every clan member is responsible for supporting the poor. Although all households are entitled to get social supports, preferences are made to female-headed households and polygamous households when they face problems. For example, a cow will be given to lactating mothers if they do not have any lactating cows.

With regard to coordinating the social support, there exist distinct forms of coordination on the basis of the gender division of labour. If the support, for instance, is related to food arrangement, women assume the responsibility of coordinating while men coordinate supports in other areas, such as collecting livestock, to help the needy ones. The FGD participants further added that they had community rotational saving group named *equip*. Membership criteria were ability to pay the contributions and willingness to abide by the rules and regulations of the groups. However, the groups have not been functioning for the last five years, as the members lack money to save due to the recurrent drought. But, FGD participants indicated that they had a plan to re-establish such groups if possible (FGD with selected elders from Aydora PA, Erer District, 17th July, 2009).

In general, these institutions are considered as traditional social protection or security mechanisms. To mention, *Zakat* which is religious alms from wealthier relatives, clan, or sub-clan members usually provided to disadvantaged groups once a year and it is mandatory.

Irmaansi is when households provide milking animals to households who have no lactating stocks across seasons to use it and return at the end of the milking period.

Maal (literally means milking) is the sharing of livestock during milking period.

Rai is the culture when children from poor pastoralists herd for richer relatives and receive food and other benefits as payment.

Keyd is the process of adopting livestock offspring from richer households, sometimes on credit. It also refers to when pastoralist women reserve the traditionally preserved butter from the special fatty cows.

Finally, *Dhowrto/Dawarsi* and alternatively named *Qaadhaa* is the contribution of animals, milk and its product for distribution amongst poor households with no milking animals, especially during dry season (*Jilaal*). Furthermore, such traditional asset redistribution systems are important social institutions in Somali community which are considered as traditional social security systems. These social systems are meant to support poor households or those who have lost many assets due to hazards such as droughts, conflicts or disease.

The Impacts of Climate Change on Women's Livelihoods and Resource-Poor Households

Climate change has a multitude of impacts on the economy (loss of livelihoods, shortage of food, shelter and water and increased poverty), on socio-political aspects (weakened states, migration due to loss of livelihoods and conflict, increased ethnic, religious and economic tensions and instability); and environmental (strong storms, droughts, water scarcity, soil erosion, desertification, deforestation and disease).

Women are more vulnerable to the effects of climate change than men primarily as they constitute the majority of the world's poor and are more dependent for their livelihood on natural resources that are threatened by climate change. Furthermore, they face social, economic and political barriers that limit their coping capacity.

Because of the limited nature of livestock mobility due to regional boundaries, there is a shift of pastoralists from their temporary houses named *dessa* to permanent houses, which necessitates the cutting down of larger and stronger trees. In addition, as they shift from a mobile to sedentary way of life their diet will also be changed to include more other foods, instead of dairy products, that need cooking. This in turn will consume more fuel wood. Furthermore, traditional institutions of governance, including those governing land and resource management, are in danger of being replaced by modern systems. Without support, there is the danger that the wealth of traditional knowledge and local experience of Somali women, together with their proven resource management, will gradually disappear.

Climate change has significant impacts on water sources, affecting the availability of water used for domestic and productive tasks. The consequences of the increased frequency in droughts are far reaching, particularly for vulnerable groups, including women who are responsible for water management at the household level. Given the changing climate, inadequate access to water and poor water quality does not only affect women, but also their responsibilities as primary care-givers, and the health of their families. It also affects the care for livestock and increases the overall amount of labour that is expended to collect, store, protect and distribute water. This means all areas, which are normally dependent on *berkad*s (traditionally constructed water ponds), are exposed to water shortage in the years where the seasonal rainfalls performs poorly and fail to refill the *berkad*s fully. *Berkad*s became increasingly important as a source of water; they are purely private owned, as anybody who can afford to construct their own in their clan territory can do so. Hence the owner of the *berkad* has full control over the use of the water.

In both the pastoral and agro-pastoral areas, climate change has negatively impacted on the physical, financial and natural assets of communities. Amongst pastoral communities, the change in water availability, in terms of timing and amount, and increases in temperature, has impacted on the rangeland and on livestock watering points. These changes, along with other risks, have resulted in a decline in the herd size per household in all livestock types and a change in herd composition. A root cause analysis indicates that the major reasons for the observed trend are lack of fodder and livestock water availability and increasing incidence of livestock diseases.

Figure 2: **Privately owned *Berkad*s in Harshen District**

Source: Photos taken by the Author

In terms of health, some potential climate change scenarios include increased morbidity and mortality due to heavy heat waves, floods, storms, fires and droughts. In addition to the impact of climate change on women's health through water scarcity and water contamination, an abundance of evidence links the evolution and distribution of infectious diseases to climate change and weather. This entails a greater incidence of infectious diseases such as cholera, malaria, and dengue fever, due to the extension of risk seasons and wider geographic distribution of disease vectors.

Conclusion and Recommendations

This section summarises the results of the research by focusing on the impacts of climate change on Somali customary institutions, and how climate change has contributed to the changing nature of pastoral livelihoods.

Conclusion

Institutions for rangeland management broke down with the introduction of modern institutions. These institutions became weak when droughts were increasingly becoming severe and frequent, and the redistribution systems were becoming impracticable, as the number of people needing social support was increasing every year.

Social associations, such as youth associations, inter-youth support groups, and women's income-generating groups are also negatively affected by drought impacts. Increased migration and longer durations spent searching for water, livestock feed and food leaves little time for social activities and meetings. Drought also impacts on savings and credit groups/cooperatives negatively, since incomes from livestock, livestock products, crops, and activities such as petty trading are highly reduced. There is an increasing rate of drop-out members from saving and credit cooperatives due to recurrent droughts. The effectiveness and sustainability of traditional resource utilisation and management systems are also threatened by increased drought frequency, due to abnormal mobility and exacerbated resources.

Climate change together with other social and environmental factors have changed pastoral livelihoods in various ways, which also has affected women's livelihoods, their coping strategies and their social position in a number of ways.

Firstly, it affects pastoral livelihoods by increasing the scale of poverty and the disruption of families particularly among the poorest groups of pastoral families.

Secondly, there is an increasing rate of pastoral mobility due to the recurrent drought and the decline of natural resources although in normal times men are more involved into far-distance movements with their herds, while women and children are left behind with the more vulnerable parts of the herds. In this case, women are left behind more frequently, but on a temporary basis. On the other hand, there is also a decrease in pastoral mobility as a response to the declining nature of the resource base that pastoralists heavily depend on due to acute shortage of water and animal diseases.

Thirdly, as a result of climate change, pastoralists' way of life is undergoing great transformation – moving from pastoral production system to a more sedentary one – as and the trend is moving towards higher vulnerability, loss of solidarity mechanisms (due to weak customary institutions) for coping with droughts, destitution and abandoning pastoral life altogether. Many households have opted for some sort of diversification in order to fill existing economic gaps caused by climate change. Such adaptation strategies include the selling of milk, petty trade, charcoal production, out-migration for wage labour, and so on, to major urban centres due to either they lost their animals and leaving livestock production or a lack of other viable livelihood options.

In general, climate change has a multitude of impacts on pastoral economy, socio-political, migration and environmental aspects. Women are more vulnerable to the effects of climate change than men. This is because of the fact that they constitute the majority of the world's poor and are more dependent for their livelihood on natural resources that are threatened by climate change. Furthermore, they face social, economic and political barriers that limit their coping capacity.

Recommendations

- The indigenous customary institutions should be revitalised and empowered to enable and rebuild the capacity of the local communities to assume greater responsibilities in the management of natural resources and decisions regarding basic services.
- Mobility of livestock is a necessity and it should be understood by the federal and regional governments and acted upon accordingly by recognising and protecting pastoralists' land and resource rights, and also ensuring that women have equal rights to men.
- Ensuring both women's and men's needs and interests are taken into account in adaptation plans is required, especially recognising the essential non–productive activities that women typically provide (water, fuel, food and care).
- There is a need to analyse of the land tenure system that distinguishes the rangeland and cropland has important policy implication for the regional governments. The land-use policy should take these variations in land tenure systems into consideration and there is also a need to improve rangeland productivity using a participatory approach.
- The specific needs of ex-pastoralists ought to be acknowledged and addressed in national and regional development strategies, given that this group is unlikely to re-enter pastoral production, the Gurgura clan in Erer District is a case in point.

References

Adger, Neil. 1999. Social vulnerability to climate change and extremes in coastal Vietnam. *World Development Series* 27, no.2, 249-269. Norwich: University of East Anglia.

Ahmed, Bashar, et al. 2008. Gender, climate change, and human security: Lessons from Bangladesh, Ghana, and Senegal. The Women's Environment and Development Organization (WEDO) with ABANTU for Development in Ghana, Action Aid Bangladesh and Environmental Development Action in the Third World (ENDA) in Senegal.

Aklilu, Amsalu and Alebachew, Adem. 2009. Assessment of climate change-induced hazards, impacts and responses in the southern lowlands of Ethiopia. *FSS Research Report* No. 4. Addis Ababa: Forum for Social Studies.

Bamlaku Tadesse, Yeneneh,Tesfaye and Fekadu,Beyene. 2010a. The role of women in indigenous conflict resolution among the Issa and Gurgura clans of Somali in Eastern Ethiopia. *African Journal on Conflict Resolution* Vol. 10, No. 1, Durban.

—.2010b. Factors affecting women's participation in indigenous conflict resolution in Issa and Gurgura clans of Somali, Eastern Ethiopia. *US-China Law Review*, Vol. 7, No. 11 Libertyville, Illinois: David Publishing Company.

Black, Richard. 2001. Environmental refugees: myth or reality? *UNHCR Working Paper* No. 34. Falmer, Brighton: University of Sussex.

Chinogwenya, Waddington & Hobsen, Matthew. 2009. Getting it right: Understanding livelihoods to *reduce the vulnerability of pastoral communities.* REGLAP paper ODI Humanitarian Policy Group, London.

Cline, William. 2007. Global warming and agriculture: Impact estimates by country. Washington: Centre for Global Development and Peterson Institute for International Economics.

Denton, Fatma. 2002. Climate change vulnerability, impacts, and adaptation: Why does gender matter? *Gender and Development,* 10(2), 10-20, New York: Routledge.

Devereux, Stephen. 2006. Vulnerable livelihoods in Somali Region, Ethiopia. Institute of Development Studies (IDS), *IDS Research Report* 57, Brighton: University of Sussex.

Diarra, Sidi. 2009. The impact of climate change on the development prospects of the least developed countries and small island developing states. Office of the High Representative for the Least Developed Countries, Landlocked Developing Countries and Small Island Developing States (UN - OHRLLS), New York: UN Headquarter.

Elizabeth, Bryan. 2008. Coping with climate variability and adapting to climate change in Kenya: Household and community strategies and determinants, Washington D.C: IFPRI,

Federal Democratic Republic of Ethiopia (FDRE). 1998. The 1994 population and housing census of Ethiopia: Results for Somali Region, Volume I – statistical report. Addis Ababa: Central Statistical Authority.

Gleditsch, Nils, P. 2010. Resource and environmental conflict: The- state-of-the-art. In Eileen Petzold-Bradley, Alexander Carius & Arpád Vincze, (eds.), *Responding to environmental conflicts: Implications for theory and practice*, Dordrecht, 53-66.

2003. Environmental conflict: Neomalthusians vs. cornucopians. In *Security and the environment in the Mediterranean: Conceptualising security and environmental conflicts*, Berlin: Springer.

Intergovernmental Panel on Climate Change (IPCC). 2001. Climate change impacts, adaptation, and vulnerability. Summary for policymakers and technical summary of the Working Group II report. Geneva.

2007. Africa. Climate change 2007: Impacts, adaptation and vulnerability. Contribution of Working Group II to the Fourth Assessment Report of the Intergovernmental Panel on Climate Change. Cambridge: Cambridge University Press, 433-467.

Jeffrey, Stark, Christine, Mataya and Kelly, Lubovich. 2009. Climate change, adaptation, and conflict: A preliminary review of the issues. *Conflict Management and Mitigation Discussion Paper* 1,14-15.

Muthukumara, Mani A. 2008. Policy and institutional reforms to support climate change adaptation and mitigation in development programs, a practical guide, Washington, D.C: The International Bank for Reconstruction and Development/The World Bank.

Myers, Norman. 1993. *Ultimate security: The environmental basis of political stability*. New York: Norton.

Odedra, Kolmannskon. 2008. *Future floods of refugees. A comment on climate change, conflict and forced migration*. Oslo: Norwegian Refugee Council.

Orindi, Victor. 2005. Mainstreaming adaptation to climate change in the development process in Uganda. *Eco_Policy Series* no. 15. Nairobi: African Centre for Technology Studies.

Oxfam International. 2008. Survival of the fittest, Pastoralism and climate change in East Africa. Oxfam Briefing Paper.

Patterson, Jacqui. 2009. Natural disasters, climate change uproot women of color. Available from http://www.ontheissuesmagazine.com/2009fall/2009fall_patterson.php. (accessed on November 21/2011)

Ryan, Laddy. 2010. Africa adaptation programme experiences, gender and climate change: Vulnerabilities and resilience in the face of climate change. *Discussion Paper Series*, Vol. 1, 4-10.

Smit, Barry and Pilifosova, Olga. 2003. From adaptation to adaptive capacity and vulnerability reduction. In *Climate change, adaptive capacity and development*. Available online: <http://www.c-ciarn.ca/index_e.asp?CaId=15&PgId=25> Accessed on July 12/2011

United Nations Development Program. (n.d.). Resource guide on gender and climate change. Available from: http://content.undp.org/go/cms-service/download/asset/?asset_id=1854911.

United Nations Women Watch. (n.d.). The UN Internet gateway on gender equality and empowerment of women. Available from <www.un.org/womenwatch. (accessed on September 2011)

USAID. 2007. Adapting to climate variability and change; guidance manual for development planning, Vols. I & II. Cambridge: Cambridge University Press.

Zelalem Tamirat and Fekadu Beyene. 2009. Governance of Erer River Small-Scale Irrigation (SSI) scheme, Institute of Pastoral and Agro-pastoral Studies, Haramaya University (unpublished research report).

Chapter 10

Traditional Adaptation Mechanisms to Climate Change and Variability among Women Pastoralists in South-west Zimbabwe

Matsa Mark and Matsa Winniefridah*

Background

Climate change refers to shifts in the mean state of the climate or its variability, persisting for an extended period, which can be decades or longer. Climate change may be due to natural changes or to persistent anthropogenic changes in the composition of the atmosphere or in land-use (USAID 2007). Humans have changed the composition of the earth's atmosphere since the pre-industrial era. The unabated use of fossil fuel currently accounts for over 80 per cent of the carbon dioxide being added into the atmosphere. Land-use and land cover changes, for example, clearing land for timber, ranching, crop farming and settlement account for between 15 – 20 per cent of current carbon dioxide emissions (Zhakata 2008).

Many of the regional variations in types of farming have a climatic origin (Bowen and Pallister 2000). Types of farming depend on weather and climate. According to World Wide Fund for Nature (WWF) (cited in Chenje 2000), there are indications of climate change in southern Africa as a result of global warming. Temperatures have risen by over 0.5°C over the past 100 years, with the 1990 - 2000 decade being the warmest (Chenje 2000; FAO

*matsam@msu.ac.zw

2004). Studies by WWF also project that southern Africa will be affected by warming in both summer and winter and that rainfall may diminish in certain areas and that soil moisture will probably decrease even more with the doubling of carbon dioxide. Southern Africa has also experienced noticeably less rainfall, and drought has become a serious threat in the region. Reduction of rainfall is significant in the Zambezi Basin countries like Botswana, South Africa, Namibia, Zambia, Tanzania, Malawi, Mozambique and Zimbabwe. Hirji, Maro and Chiuta (2002) note that climate change and variability has always been a fact of life in southern Africa and that this variability is expected to increase. Variability is expected to occur over the interior upland of the Zambezi Basin encompassing Botswana, eastern Zambia and western Zimbabwe.

In southern Africa both subsistence and commercial agriculture depend on rainfall which is variable and unreliable across regions. In addition to variability between countries, rainfall is also variable within a country across seasons. Recurrent drought and floods have resulted in severe and localised shortages of the main cereal crops and other foods leading to dependence on traditional and other economic activities. In Zimbabwe's south-western provinces of Matabeleland South, Matabeleland North and parts of Masvingo pastoralism has thus remained the backbone of household wealth and livelihoods (Hargreaves, Bruce and Beffa 2004).

Implications of climate change for Zimbabwe are serious. The number of years with below average rainfall is increasing. According to Zhakata (2008), there are signs of a gradual warming in both summer and winter temperatures and models suggest that south-west Zimbabwe's largest city, Bulawayo, will experience a minimum rise in daily temperatures of 2.7 °C before the end of the 21st century. A survey on farmers' perceptions of climate change in Zimbabwe by Masendeke (2008) indicates that farmers have noticed changes in the quantity, quality and efficacy of rainfall. There is a general decline in the amount of rainfall, which is more pronounced in the semi-arid tropics, the largest part of which is constituted by south-west Zimbabwe – the study area. In terms of rainfall quality, farmers reported differences in distribution as well as more erratic rainfall events.

The dry spells appear to be increasing in duration and frequency and this is also more pronounced in semi-arid tropics. There is an increased incidence of drought and farmers noticed that rainfall is less effective now than before. Rainfall has become less frequent, causing more run-off and soil erosion. Season duration has changed. The rainy season used to end in March/April but currently it ends as early as February. The country's first late maturing maize hybrid SR52, which used to be grown in the south-western districts of Tsholotsho and Gwanda when it was first introduced in the 1960s, can no longer be grown successfully in these areas.

Farmers have also noticed environmental changes which include the drying up of wetlands, ponds, pans and riverbeds. Certain grass species that used to be found in wetlands are disappearing together with related small insect species. Farmers conceded that they can no longer rely on their traditional knowledge or signs, for example, the flowering of certain tree species, or a halo around the moon or sun which used to signify rain (Masendeke 2008).

Climate variability has always been a fact of life in southern Africa and the emerging challenge is the expected increase in variability due to global climate change. Climate variability refers to variations in the state of climate on all temporal and spatial scales beyond that of individual weather events. Examples of climate variability include extended droughts, floods, and conditions that result from periodic *El Niño* and *La Niña* events (USAID 2007). Recurrent droughts and floods that southern Africa has experienced have resulted in the loss of human life, livestock and other property as well as severe localised shortages of the main cereal crops and other food. During the 1991 - 92 droughts in Zimbabwe, average annual precipitation fluctuated from 335 - 1004mm and averaged 640mm (Hirji, Maro and Chiuta 2002). It is projected that in southern Africa rain may diminish in certain areas. Reduction is significant in the Zambezi Basin, which includes Zimbabwe and Botswana where livestock production is one of the major sectors in agriculture (Chenje 2000). Success of the livestock industry depends much more on the availability of pasture that in turn depends on availability of water resources.

Both subsistence and commercial agriculture in the Basin depends on rainfall which is variable and unreliable across the region. In addition to variability between countries, rainfall is also variable within a given country across seasons, thus making rain-fed agriculture risky.

Adaptation refers to any adjustment, passive, reactive or anticipatory, that is proposed as a means for ameliorating the anticipated adverse effects associated with climate change (Stakhiv 1993, quoted in Smit *et al.* 2000). It involves adjustments to enhance the viability of social and economic activities and to reduce their vulnerability to climate, including its current variability and extreme events as well as longer term climate change (Smit et al. 2000). The Inter-governmental Panel on Climate Change (IPCC) (2006, 1) defines climate change adaptation as "adjustments in ecological, social or economic systems in response to actual or expected climatic stimuli and their effects or impacts". Climate change adaptation, therefore, involves changes in processes, practices and structures to moderate potential damages to or benefit from opportunities associated with climate change.

Vulnerability is the extent to which a natural or social system is susceptible to sustaining damage from climate change (IPCC 2006). It is the degree to which a system is susceptible or unable to cope with adverse effects of climate change, including climate variability and extremes (Olmos 2001). Climate change is, therefore, a function of the character, magnitude and rate of climate variation to which a system is exposed, its sensitivity, and its adaptive capacity.

Vulnerability can also refer to the ability or inability of individuals or social groupings to respond to, recover from or adapt to any external stress placed on their livelihoods and well-being (Kelly and Adger 2000). This definition is more relevant in this discussion since it defines vulnerability as being determined by access to resources and specifically by individuals' entitlement to own these resources.

Gender issues have been rarely addressed, particularly in relation to climate change impact in global climate change initiatives and policy meetings (UNDP 2009). Debates have focussed mainly on mitigation efforts required by governments and institutions to mitigate climate change. Insufficient attention has been given to climate change adaptation strategies, that is, actions taken for and by people to prepare and make changes to protect themselves against climate change. Women's roles are not adequately recognised or accounted for in climate change mitigation and adaptation efforts, in national and global climate negotiation talks or in the context of natural disasters influenced or exacerbated by climate change.

The vulnerability of women to the effects of climate change is mainly because of their over-dependence on natural resources, their responsibility for water and food procurement and their increased risk exposure during times of disasters and severe weather crises (UNDP 2009).

According to FAO (1995), pastoralism is a branch of agriculture concerned with rearing of livestock such as goats, cattle, sheep and camels. The Department of Rural Economy and Agriculture (DREA) (2010) describes it as a way of life characterised by high reliance on livestock ruminants as a source of economic and social well-being in areas of high rainfall variability. It is animal husbandry that also relies on use of extensive grazing or rangelands for livestock production. Pastoralism is practised in many countries of the world. In Africa, it is found in all regions, mostly in vast, arid and semi-arid areas characterised by rainfall variability and uncertainties in the spatial and temporal distribution of water and grazing land for animals (*Ibid.*). Rohrbach (1999) points out that pastoralism in Africa occupies 40 per cent of Africa's landmass although with variations between countries. The success of pastoralism depends on population density and the presence of extensive grazing lands and water in these regions. Pastoralism has, however, changed forms and many of its traditional practices have adapted to the changing environments and circumstances of the modern world (Ellis 1995; FAO 1995).

Southwest Zimbabwe is an agro-pastoral community with rain-fed cultivation and animal husbandry as major economic activities. It is a marginal area that experiences recurrent droughts and floods which negatively affect food security and food self-sufficiency. The average annual rainfall fluctuates from 335 to 1004mm (between 1970 and 2000) and averages 640mm. Staple maize production is strongly influenced by these climatic factors. Such fluctuations affect the livelihood of these agro-pastoralists. They therefore rely heavily on livestock in times of the recurrent droughts and famine. Other economic activities, thus, revolve around pastoralism; for example, cultivation depends on livestock (draught power) and manure. Livestock also provide transport, skins for yoking, meat and milk. Livestock may also be exchanged for cash, furniture, groceries, blankets, school fees, seeds, building material and other modern goods like bicycles and scotch-carts. The pastoralist burden in this region, however, falls heavily on women, although because of the patrilineal nature of society their efforts are never hailed, with credit going the way of absentee husbands. Most of the able-bodied men in south-western Zimbabwe are migrant workers in neighbouring South Africa and Botswana and women are the *de facto* heads of households taking care of families in all respects. Their success in fending for families in the face of climate change and variability to this day is an achievement worth investigating and highlighting. This paper, therefore, assesses the role of these pastoral women in coping with and adapting to climate change and variability. It traces the historical and traditional context of pastoral women's existence in south-western Zimbabwe. The paper also assesses the role played by pastoral women and explores their coping and adaptation mechanisms to climate change and variability. It finally identifies pastoral women's constraints in their attempt to cope and adapt to climate change and variability in south-western Zimbabwe.

The few studies on climate change and variability in southern Africa tend to be gender-inclusive and rarely address the experiences of pastoral women as narrated by them. This study attempts to bridge that gap. Research also tends to highlight the harmful effects and top-down mitigatory measures of climate change and variability without looking at pastoral women's adaptation mechanisms. Climate change and variability problems are usually addressed at national (macro-scale) level rather than at local (micro-scale) level. This leads to ineffective solutions especially for disadvantaged groups like pastoral women. Climate change is a process which pastoral women live with and have experienced over years, yet problems of climate change are normally only addressed after a significant event like a severe drought, floods, and storm, among others.

This study, therefore, informs government and other stakeholders on the real issues affecting women in marginal pastoral communities. It can help stakeholders who might be interested in designing projects and programmes which are specific to helping out pastoral women in specific communities. The study also informs stakeholders of community adaptation mechanisms tried and tested over generations so that they are blended with modern initiatives for robust results. Ultimately it is hoped that the study will impress upon stakeholders to appreciate the need to preserve local community resources upon which communities depend and hold sacrosanct.

Literature Review

Pastoralism in Zimbabwe: A Historical Overview

Historically, pastoralism originated as nomadic where there was heavy reliance on intensive management of livestock herds. This was consolidated by the hunting of wild animals and gathering of wild fruits. Notable pastoral groups in southern Africa were the San and the Khoikhoi (Davis 1978; Bhila and Shillington 1986; Prew *et al.* 1993). Hargreaves, Bruce and Beffa (2004) also point out that in 1900 pastoral systems were the norm and there was little integration of livestock with crop farming. Pastoral nomadism is still significant in African semi-humid tropics where conditions are unfavourable for crop growing. Mobility allows the groups to exploit more than one environment, thus the system always moves to suit different and unpredictable environments. With domestication of animals and a settled way of life, migration of heads becomes seasonal or daily. Sometimes households move with seasonal migration. The actions of herders are carefully planned and adjusted to match the changing conditions of the environment. However, pure nomadism no longer exists in southern Africa (Carr 1997).There has been a shift to agro-pastoralism for self-sufficiency in food security. The Department of Rural Economy and Agriculture (2010) also notes that different types of movement depend on the local environment and type of livestock reared. The movements are rational, highly strategic and draw from local information and risk analysis.

The colonial legacy changed the role and the influence of traditional pastoral institutions and created restricted boundaries that divided well established social and economic units (Prew *et al.* 1993; DREA 2010). During colonialism there was privatisation of land, and as pastoralist populations soared, pastoralist mobility was restricted (Carr1997). Pastoralists were coerced into adopting a more settled way of life on the pretext of trying to provide sanitation and medical care (Tevera and Moyo 2000). This disturbed the previous balance of the pastoral system. In South Africa, Botswana and south-west Zimbabwe, the nomadic San and the Khoikhoi were forced into settled way of life (Davies 1978; Carr 1997; Tevera and Moyo 2000).

The Land Apportionment Act of 1931 in Zimbabwe divided the land into three land tenure systems, namely: commercial farms for settler farmers; land purchase areas for the educated black elite; and communal areas for the majority black population (Prew *et al.* 1993). Each household in communal reserves had a well defined cropping area and access to communal grazing. Woolmer, Sithole, and Makamuri (2002, cited in Hargreaves, Bruce and Beffa 2004) point out that in the 1940s Agricultural Extension System began to promote mixed farming. The cattle tillage system where farmers used oxen for ploughing was introduced. As populations expanded, there was a need to intensify food production on smaller areas through the use of draught power and manure from cattle as fertiliser. Land for grazing was reduced. The colonial administration allowed co-existence of traditional authority and power structures within the Western style of administration. Pastoralism remained in ancestral communal lands where resources were owned communally. The British indirect rule allowed the existence of powerful chiefdoms and chiefs to rule pastoral communities. These distributed land and resolved land conflicts in this new tenure system (Prew *et al.* 1993; DREA 2010).

The privatisation of land for commercial farms and African Purchase Areas had negative impacts on availability of land for grazing in Communal Reserves. Pastoral lands were appropriated. Africans were driven out of good land to reserves with poor soils and little rainfall; for example, Gwai, Shangani, Lupane, Tsholotsho and Nkayi were cases in point. Mobility was restricted because it was viewed as irrational, inefficient and causing land degradation. This was followed by the Land Husbandry Act of 1951 which reduced the number of cattle per household (Prew *et al.* 1993). This, coupled with cattle tax, dipping fees and fees for grazing lands, reduced the number of cattle. Payment for taxes and fines was also in form of cattle. Transfer of land from tribal people to individuals and state, and taxes disturbed the balance of patriarchal systems.

The decolonisation of the African continent was followed by modernisation which meant a shift from subsistence to commercial agriculture, privatisation of pastoral land, and thus changes from traditional pastoral management systems to modern ones (DREA 2010). Conroy (2002) also asserts that in post-colonial societies, common property rights (CPRs) have come under increasing pressure in recent decades with the growth of human and livestock production and trends towards privatisation of land. Many traditional management systems have broken down. Land- and water-based CPRs have become degraded and are shrinking. Since the last 10 - 20 years, inter-development agencies and governments have been trying to rehabilitate CPRs and to improve people's access to them by promoting shared or joint management regimes. One example is the Zimbabwe Land Reform Model A1 scheme for resettlement promoted ownership of common grazing lands and communal access to water.

Importance of Pastoralism in South-West Zimbabwe

The Department of Rural Economy and Agriculture (2010) reveals that in general pastoralism contributes to 10 – 44 per cent of the Gross Domestic Product (GDP) of African countries. However, it notes that statistics overlook many important economic benefits of pastoral livestock products, such as milk for children and pregnant mothers, manure, transport, draught power, fuel floors, and treatment for grain and skins. Livestock is also an insurance against shocks like crop failure and during religious memorial, spiritual appeasement and customary rites as well as festive occasions. Van Rooyen and Homann (2009) assert that in southern and western Zimbabwe, the Matabele play an important role in the livelihoods of small-holder farmers. Goats are valued primarily for meat, milk, manure, skins and cash for petty commodities. Barret (1992, cited in Hargreaves, Bruce, and Beffa 2004) study in Zimbabwe reveals that in Zvishavane in 1986 - 1988 the estimated economic value of livestock stood at 57 per cent draught power, 22 per cent milk, 16 per cent transport and 5 per cent manure, sale and slaughter. He adds that cattle production is related to crop production. Cattle provide inputs such as draught power, manure, transport and seeds through barter exchange or for cash. 90 per cent of communal tillage is done using livestock (Hargreaves, Bruce, and Beffa 2004). Although the value of donkeys is under-estimated, Nengomasha *et al.* (2007) states that donkeys are increasingly becoming an important mode of transport and an alternative source of draught power because they are cheaper, easy to train and less susceptible to diseases. Donkeys also survive on little water, poor quality food and lower food requirements than cattle. Donkeys also have less gender restrictions relating to use, have longer working life than cattle and do not need roads to reach remote parts of the communal areas, especially in newly resettled areas of Zimbabwe.

Gender, Climate Change, Pastoralism and Adaptation

The majority of pastoral societies are patrilineal (Ellis 1995). As a result, ownership of livestock is influenced by gender. Men dominate in ownership, acquisition, use and disposal (Chenje 2000; Hargreaves, Bruce, and Beffa 2004).

Kraub (2011) notes that social consequences of climate change are not gender-neutral. Men and women are affected differently by climate change. Gender differences reduce the survival chances of men and women during climate-related natural disasters like water shortages. In response to climate change, gender adaptation and coping strategies differ. Men and women experience different vulnerabilities and play different roles in coping to natural disasters (Cannon 2002). Women's informal rights to resources (land and water), patriarchy, (which has a bearing on women's autonomy relating to crop and livestock), lack of finance and lack of knowledge (resulting from low levels of education), determine women's adaptation and coping strategies (Mogotsi, Nyangito and Nyariki 2011). Women tend to rely on traditional coping strategies because modern ones are not readily available to them. For example, in response to shortage of rangelands, women move long distances for grazing and gather wild fruits, grass fodder, tree fodder to feed cattle and small ruminants of goats and sheep (Conroy 2002).

Women also depend on surface water bodies and for weather forecast, they rely on traditional signs like flowering of certain tree species, fruits, halo around the sun or moon or patterns of winds (ICRISAT 2008).To this note, Cannon (2002) deduces that women's knowledge of bio-diversity relating to conservation and domestication of plants and animals is key to adapting to climate change.

Rohrbach *et al.* (2004) and Hargreaves, Bruce, and Beffa (2004) classify adaptation into planned and unplanned adaptation. Planned adaptation has mitigation measures, mostly modern, and is spearheaded by governments and non-governmental organisations (NGOs) through agricultural extension services and veterinary services. Men tend to respond more to planned adaptation than women because of better access to information, financial strength and access to education (formal and informal). Coping strategies under planned adaptation include cattle diseases surveillance, vaccinations, transfer of livestock technology to small-holder farmers who are mostly males, loan schemes, cross-breeding heads, water management technologies, distribution of stock feeds and drilling boreholes. During the 2002 - 2003 drought in Zimbabwe, 200 tonnes of stock feed were distributed to Beitbridge, Matobo, Kezi and Bulilima districts.

Unplanned adaptation is usually based on traditional knowledge systems and is usually associated with women because the strategies are affordable. The coping strategies include destocking, relocating livestock to better pastures, cutting hay, digging up and chopping tubers for livestock, harvesting fleshy creepers from rocky outcrops, utilising tree fruits, barks and branches as fodder sources, digging wells in perennial rivers, preserving harvest stock from the fields and using traditional methods of treating livestock (Hargreaves, Bruce, and Beffa 2004, cited in ICRISAT 2008).

Policy Framework on Pastoralism in Africa

The Department of Rural Economy and Agriculture (2010) notes that it saw it necessary for African states to come up with a policy framework on pastoralism to secure, protect and improve livelihoods and rights of pastoral communities. It is noted that pastoralist culture was part of cultural heritage which contributes significantly to the national and regional economies of Africa although it is under-valued. It is viewed as necessary to have a policy framework that emphasises full involvement of pastoralists' men and women in the regional development processes from which they are supposed to benefit. The aim of the policy framework is to develop appropriate pastoral policies relating to access to rangelands, market of pastoral livestock and their products and to fully integrate pastoralism into regional development programmes and plans. There is need to recognise the rights and existing economic contribution and potential future contributions of pastoralism to development (DREA 2010).

It is realised that pastoralists represent a substantial segment of the population in many African countries although they are not represented in governance. Policy developments are also to embrace indigenous knowledge and innovations for sustainable natural resources management. Pastoral needs are to be reflected in national policies and planning, for example, the needs that relate to infrastructure, water resources, education, health provision, human capital and market facilities.

These also include management of risks to reduce vulnerability to climate eventualities like droughts and floods. The other objective is to lay strategies to manage conflicts resulting from conflicts over rangelands and water sources (common property pools). Another major aim is to enhance greater participation of pastoral communities in decision-making and policy debates (*Ibid.*).

Theoretical Framework

This paper uses Vandana Shiva's eco-feminist perspective which explores women's intimate knowledge of the environment (Momsen 2004). It also uses feminist geography as a tool for investigating the role played by women in male-dominated social structures. Adaptation theories which posit that human societies are not passive but respond and adapt to changing environments (Schaefer 2006) also form part of the tools of analysis for this paper.

Ford (1983) notes that the process of adaptation is a dynamic process that takes place over generations. New adaptation strategies are needed when the organism encounter a new environment. Adaptation occurs in response to changes in the environment and life styles. Organisms that are not suitably adapted to their environment will either move out of the habitat or die out. Adaptation can be structural, physiological or behavioural. Our concern with pastoralists as human beings is behavioural adaptation. Settel (1999) argues that humans adapt to their environmental conditions through change of behaviour and actions as they interact with their natural environments. They design new adaptation and coping strategies to survive in their environments.

Shiva (2011) points out that natural disasters cause disharmony with nature leading to non-sustainability. This leads to violation of the laws of the earth and nature, hence violating humanity. She argues that harmony with nature is imperative. The Millennium Development Goals (MDGs) relating to food and water cannot be met if people are not in harmony with nature. She further points out that the right to Mother Earth is crucial for meeting the needs of people of different races, genders, religions and ethnic groups.

Eco-feminists, for example, Vandana Shiva, Carolyn Merchant and Susan Griffin, believe that there is a strong connection between women, human rights and the exploitation of nature (Chandler and Wayne 2002).

Merchant (1995) says women have a central role in preserving and understanding nature. Women need to be included in the construction of environmental knowledge. It is unfortunate that patriarchal power has made women to appear as if they have turned against the environment instead of living in harmony with it. Women have little access and control of the natural environment, and they are discriminated against in the dissemination of information relating to the environment. Their indigenous knowledge (IK) is looked down upon and is viewed as unscientific (Njiro 1999).

Kraub (2011) also points out that natural disasters and climate change affect males and females differently. Women lack knowledge to adapt to climate change situations because of lack of information, lower levels of education, cultural and patriarchal attitudes that limit their opportunities. They also lack financial and legal framework to pursue alternative livelihoods. There is, therefore, a need for integration of women into climatic change issues and management of natural and man-made environmental issues for both males and females to live in harmony with the environment.

Feminist geographers view geography, which is the study of "man and his physical environment" as entirely male-dominated (Boyer 2004). Women exist as hidden from geography. Little effort is made to investigate the role they play in society other than adjusting to a male-dominated and determined order. Feminist geography takes into account the socially created gender structures of society and is committed towards alleviation of gender inequality through social change. Women are also less free to adapt to new conditions because they face discrimination in a number of ways. Women are also restricted to reproductive roles and have to cope with family obligations, making them less mobile than men. Modernisation is thus viewed as trivialising the role of women through cultural systems, property holding, new technology and education.

In pastoralism, women's role is trivialised despite the fact that they are the ones who carry both the burden of reproductive roles and full time pastoral activities. They are less mobile when it comes to migration for employment (Matsa and Matsa 2010). They are also side-lined as far as technology, land and livestock rights, and access to knowledge and information relating to pastoralism and environmental management and adaptation strategies are concerned. As a result, Boyer (2004) notes that geography and gender should be reconsidered as a way of combating gender inequalities in the study of "man and his environment".

Methodology

Description of the Study Area

South-west Zimbabwe covers about one third of Zimbabwe and is constituted by 18 districts (see Figure 1). To the south it borders with South Africa and Botswana while to the west it borders with Namibia and Zambia. It is economically a marginal region and shortage of water is its main problem. The greater part of south-west Zimbabwe is an extensive farming region. Rainfall in this region is too erratic for reliable production of even drought-tolerant fodder or grain crops. As a result of this, farming is mainly restricted to veldt utilisation, where cattle ranching is the only sound farming system for this region. Most areas in this region are below 900 metres altitude and mean rainfall is between 500 - 700mm with a coefficient of variation of around 30 – 40 per cent, while mean annual temperature is between 20 °C and 22 °C (Chenje, Sola and Paleczny 1998).

The main soil types in south-west Zimbabwe are regosols derived from the Kalahari geological sand system which extend into Botswana. The soils have low nutrient reserves. They are permeable and have a low moisture retention capacity and this limits crop production. Sorghum, millet and some drought-tolerant maize varieties, pumpkins, and water melons are the most common crop types grown in the region.

Figure 1: **Map of south-western districts of Zimbabwe**

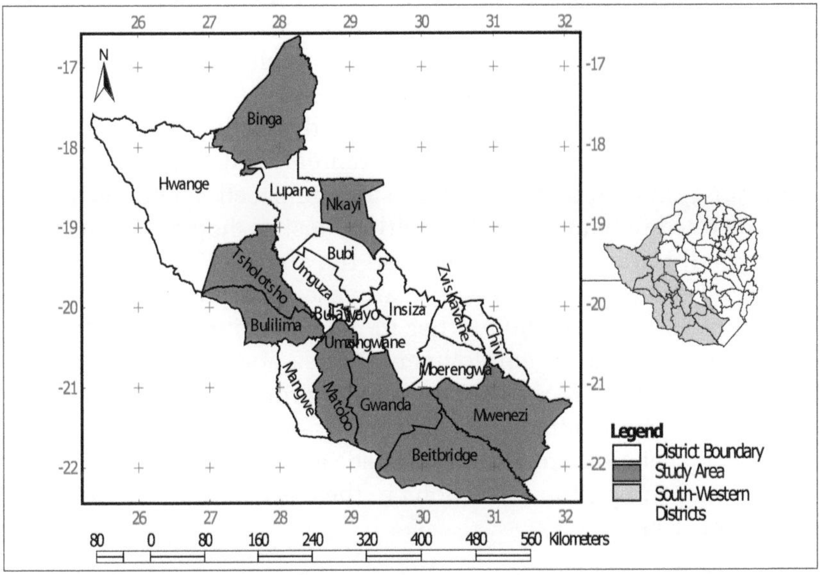

Source: GIS-Generated by the authors

On heavier soils along river valleys, *brachystegia boehmii* and *brachystegia speciformis* tree species dominate. In the driest parts of the Kalahari sandveld like Tsholotsho, Bulilima and Mangwe districts, *colophospermum mopane* low woodland is locally common on heavier-textured soils and in areas of restricted damage. *Combretum collinum* open woodland is characteristic of the Karoo sandstone soil of Jotsholo and Kezi. Pans are common and support secondary moist grassland. Several love grass species are characteristic and these have influenced pastoralism. The northern bounding limits of pastoralism in the north-western part of the region had over the years been checked by the presence of the *trypanosomiasis* vector (tsetse fly) which has since been eradicated. To the south, pastoralism has been limited by inadequate water supplies.

Nguni and Tuli are the main types of cattle found in south-west Zimbabwe where they strive in drier environments. They are hardy, resistant to ticks and are efficient converters of poor quality feed to draught power, milk and meat (Hargreaves, Bruce, and Beffa 2004). Matabele goats are the main indigenous breeds in the region. These are hardy and highly fertile with multiple births (*Ibid.*).Sheep are the short- and fat-tailed and resemble the Chinese hard pig.

The region has limited economic activity and the regional capital, Bulawayo, which was the country's industrial hub at independence in 1980, has been on a steady decline due to multiple factors but mainly because of perennial water problems and paralysing land reform-related western imposed sanctions which have resulted in many industries either closing shop or relocating to the capital, Harare. This has reduced the already limited prospects for employment and as a result many economically active people, mainly males, have drifted either to neighbouring South Africa or Botswana in search of employment (Matsa and Matsa 2010). This labour migration has resulted in a large proportion of households in the region being female-headed. In Gwanda, for example, 50 per cent of households are female-headed, and women make most of the farming and pastoral decisions in the absence of their husbands (Rohrbach 1999).

Research Design and Sampling Procedures

The study was conducted between May and September 2011. Qualitative research design was used as it enabled researchers to obtain in-depth data concerning pastoralist women's views, experiences, feelings and behaviour as they responded to climate change and climate variability issues. South-western Zimbabwe comprises part of Masvingo Province (3 districts), part of Matabeleland North Province (7 districts) and the whole of Matabeleland South Province (6 districts). Districts for this study were selected using purposive sampling through which the most representative pastoral districts for each province were selected. Thus, for Masvingo Province Mwenezi District was selected. For Matabeleland North, Tsholotsho, Binga and Nkayi districts were selected while for Matabeleland South Province, Bulilima, Matobo, Gwanda and Beitbridge districts were chosen. Stratified sampling was used to select 3 males and 6 females from each of the 8 districts (giving a total of 9 people per district). This was done in order to capture the female voice more in line with the demands of the study and also given that female-headed households predominate in all the districts chosen (Matsa and Matsa 2010). One respondent from each household was purposively selected from 72 households in all 8 districts giving a total of 24 males and 48 females. The basis for purposive sampling was age, whereby only those household heads or their representatives 40 years or/and above, had lived in sampled areas for more than 15 years and have pastoral experience were selected. Preference was given to elderly people who were believed to have knowledge and experience covering several historical periods of pastoral existence.

Data Collection and Analysis Methods

In-depth interviews were used as the main instrument for data collection in this research. One of the researchers covered the 4 south-eastern districts of Mwenezi, Beitbridge, Gwanda and Matobo while the other covered the north-western districts of Bulilima, Tsholotsho, Nkayi and Binga. Questions sought insights in four main areas, namely: the historical and traditional context of pastoral women existence; women's roles in pastoral communities; constraints encountered by pastoral women; and finally the strategies used by pastoral women in south-west Zimbabwe to cope with and adapt to climate change and variability.

Direct observation was also used to gather data. The observation checklist used included items like availability and sustainability of water sources for livestock; availability and adequacy of rangelands; coping of households with muddy kraals in times of floods and incessant rain; preparation and pilling of stock feeds (for example, grain stalks, melons, grass, wild fruits) for posterity; livestock absence to distant rangelands (*imilaga*) due to pasture depletion nearer home; and livestock breeds in relation to the climatic region under study. Observation was thus meant to provide first-hand information about pastoral communities' adaptation mechanisms.

Focused group discussions (FGDs) were also used to collect data for this study. Researchers initiated FGDs with women at water sources where they take turns to fetch water either for domestic use or for livestock. Similar discussions were held at community development (food-for-work) projects, food distribution points by government, non-governmental departments and non-governmental organisations. Findings from this technique were used to buttress results from the other two data collection techniques already discussed. Focussed group discussions enabled researchers to gather a collective 'woman's voice' concerning women's pastoral experiences in relation to climate change.

The instruments used, however, had their own limitations, for example, when conducted at household level, it was observed that women were bound by invisible patriarchal control where, instead of speaking 'the experienced', they instead spoke 'the expected'. In other words, women spoke what is expected of them as women in the home and not necessarily their true objective experiences. This weakness was, however, counteracted by FGDs where women were removed from the household and thus had more freedom of expression. The weakness of observation as a data collection tool was that the period during which the research was conducted denied the researchers the opportunity to observe some activities which are season-determined, for example, shifting of kraals due to floods and incessant rain. To counter this, researchers depended solely on interviews.

Descriptive analysis was used in this research to analyse data from interviews, direct observations and focus group discussions. The results are presented in a narrative form. Some responses from in-depth interviews were, however, captured and presented verbatim.

Data Presentation and Analysis

The History of Pastoralism in South-West Zimbabwe

Many of the study participants revealed that before colonisation by the British in 1890 pastoral families had heads of cattle ranging from 15 to over 150. Reasons given were that the population density was low, settlements were few and rangeland resources abundant. Rainfall regimes then were more reliable with very little variability and this resulted in abundant water sources like paddles, ponds and most rivers were perennial.

Cattle had unlimited movement and each household had its own traditional livestock mark (*uphawu*). This was placed on the ears of its head of cattle for easier and quicker identification. During times of drought cattle were driven to far off places which had virgin rangelands (*imilaga*) which during normal years were not utilised or grazed.

Then, male pastoralists would go and build temporary shelter in these rangelands until the onset of the rains when grass would recover in the traditional rangelands nearer home. Types of cattle then were the indigenous *Tuli* and *Amabulu* breeds which were resistant to drought and diseases.

Cattle rustling was not common because cattle were not a viable commercial commodity then and community relations were very strong and rules very strict. Traditional leaders had uncontested power and authority to discipline wrongdoers and even their families.

Madlela (84) from Gwanda hinted that royal families had more cattle than ordinary households. However, because of these large numbers, they could not keep all these cattle in the royal homestead, hence, they seconded sizeable heads to poorer families among their subjects (*ukususisa*). This benevolent practice meant that all households had cattle which they could use at least for farming, milking, manure and other such services. They, however, had no control over these cattle in terms of selling them, killing them or bestowing family rituals on them.

During this period, Mathwala (80) from Matobo alluded to the fact that pastoralism was male-dominated. Families settled in extended family lineages with strong family bonds. All able-bodied males lived together with their families since there were no towns, mines or commercial farms to absorb them. There were also no schools meaning that all children were raised and remained in the home. Gender roles were, therefore, well defined. Livestock management was a strictly male domain. Even in families where there were no boys, male relatives from the extended family would look after their livestock. The role of women in livestock management was confined to preparation and processing of livestock products like meat and milk. Mabhena (74) from Binga and Magadlela (58) from Matobo pointed out that there were certain traditional medicines used which were believed to protect livestock from being bewitched, stolen or struck by lightning. Cattle could also be conditioned by means of some traditional treatment to come back home on their own from rangelands at the end of each day. A certain tree plant species *intakubomvu,* for example, was given to each heifer to enable it to achieve annual productivity once it started bearing calves. This traditional treatment of the head (*umthuso*) was associated with taboos, most of which distanced women from livestock management. Gogo Mabena (78) from Binga, for example, revealed that if women involved themselves in livestock management, the treatment would be defiled endangering the whole head.

According to Ngwenya (74) and Mathwala (80) both from Matobo, the advent of colonialism disturbed the harmonious kinship-based extended family structure of pastoral communities. British colonialists introduced boundaries through the Land Apportionment Act of 1930, demarcating towns, mining claims, farms, game reserves, national parks and conservancies. Pastoralist communities did not participate in this demarcation exercise and the new boundaries did not respect pastoralist way of life, beliefs and traditions. Madlela (84) from Gwanda explained that the land apportionment meant that Africans who had occupied areas which became part of the large-scale commercial farms and African purchase areas were driven to what became known as tribal trust lands where they were forced to crowd. Others, like most of the pastoral communities in south-west Zimbabwe, were driven to what became known as African reserves (*Amagusu*).These were very hot, dry, rocky areas with deep sandy soils and thick woodlands and dangerous wild animals like lions, hyenas, buffaloes, leopards, jackals, baboons and elephants, most of which are livestock predators. They were also infested with pests like tsetse-flies and mosquitoes (for example, Binga, parts of Nkayi and Tsholotsho) and diseases like nagana and sleeping sickness which attack both people and animals.These areas also had poisonous plants, like *umkhawuzane,* which kill cattle upon consumption. People from parts of Gwanda, Fort Rixon and Matobo were moved to Gomoza, via Mabhiga across Shangani River and settled in present day Lupane, Nkayi, Binga and Tsholotsho.

Upon arrival in these areas, most of their livestock died from these unfavourable conditions and diseases. Some people also died.

Two major aims of land apportionment, according to Ngwenya (74) from Matobo, were:

Firstly, to ensure that white colonialists occupied and owned the best land in terms of soils, mineral wealth, water resources and even plant and animal resources. The areas they occupied also had the best climate in terms of rainfall and temperature regimes.

Secondly, the colonialists wanted to ensure that the native blacks find life in their traditional rural communities very uncomfortable so that they would be easily forced into towns, farms and mines where they would provide cheap labour for their newly established capitalist economy. This also helped them raise the mandatory cattle tax and hut tax. To ensure the success of this colonial grand plan, more laws relating to land and livestock ownership and management were introduced. The Land Husbandry Act of 1951 reduced the number of cattle for each African household to not more than 10 for ordinary people although royal households were allowed an unlimited number. However, Ngwenya (74) and Magadlela (58) of Matobo revealed that cattle tax of 25 shillings per livestock was too high even for these royal households. The fact that colonial authorities did not accept the equivalent of this amount in any other form (not even cattle themselves) except cash made it even more difficult for Africans to own more cattle. Ngwenya and Magadlela further conceded that if cattle for a given household were more than the stipulated 10 head, they either had to sell to the white commercial farmers or lend the excess head of cattle to other trustworthy people in the community who did not have cattle. As part of destocking, white colonialists seized cattle from Ndebele pastoral communities claiming these were King's cattle that had been seconded to them by the deposed King Lobengula. To ensure that each household paid tax, the colonial regime introduced the stock card registered in the name of the male household head despite the fact that some of the cattle like the matrimonial bride price cow (*inkomo yohlanga*) specifically belonged to women.

According to Madlela (84) the new rules, regulations and laws related to livestock ownership and management hit hardest on the socio-cultural and economic harmony of pastoral communities. To fend for their families, men were forced by the prevailing unbearable circumstances to leave their pastoral niche and went to provide cheap labour in towns, mines and commercial farms. Schools were introduced and these absorbed the economically active children, especially boys who were preferred for schooling over female children. This period, therefore, signalled the shift of livestock management and use from males to females although ownership and decision-making over livestock remained a prerogative of the absentee male household head. Manyoni (70) from Bulilima reminisces that the armed liberation war period 1972 - 1979 was a difficult time as the Rhodesian army poisoned pastures and rangeland water sources as part of their war strategy to frustrate the pastoralists whom they accused of feeding the "terrorists". Many cattle died as a result.

Despite many negatives, some respondents said colonialism also had some positive aspects to pastoral communities. Colonialism introduced and enforced treatment of livestock to control the spread of diseases, for example, dipping to control ticks as well as vaccines to control foot-and-mouth diseases. Before the introduction of dipping, pastoralists relied on stock bird (*amalanda*) picking ticks from their livestock which was unreliable. Madlela (84) from Gwanda, however, contested dipping saying that it was accompanied by a dipping fee of six pence per beast and later one shilling per beast. This was expensive. Paddocking also controlled movement of livestock and this limited the spread of diseases and controlled grazing of the rangeland. It also reduced thefts and cattle stray.

Colonialism, according to Ngwenya (84) from Matobo, also introduced household-specific livestock brands which improved cattle identification and reduced thefts. Dehorning reduced the risk of cattle fatally goring one another, it also improved cattle quality. As consolation for displacement and to protect their own cattle from possible infection, the colonial government built earth dams on perennial rivers and also drilled wells for the new settlements to get reliable sources of water.

Ndlovu (50) from Gwanda conceded that white commercial ranchers introduced cross-breeds between the indigenous *Tuli* and *Amabula* on the one hand and their exotic Brahman cows (*amanjanja*) which improved both cattle sizes and their growth rate. Some respondents, like Mafu (70) from Matobo, however, complained that the cross-breeds gave less milk and were less resistant to diseases. This view was supported by Sebata (65) from Beitbridge.

Gogo Mabhena (78) from Binga observed that the advent of independence in 1980 led to a relaxation in the implementation of most of the rules, regulations and laws relating to the Land Husbandry Act. Cattle tax and hut tax were suspended. De-stocking stopped, meaning that households could own more cattle than the hitherto stipulated 10 per household. The deliberate killing of livestock through poisoning of pastures and rangeland water sources stopped. The land tenure system, however, did not change significantly because the willing buyer-willing-seller mode of land reform did not succeed. Pastures, therefore, remained the same. The number of cattle per household did not register a sharp increase due to increased recurrent droughts which stifled the regeneration of pastures. Ndlovu (60) from Mwenezi observed that 1981-1983 were serious drought seasons, with the 1994 - 1995 periods being the worst drought in living memory according to Madlela (84) from Gwanda. These droughts seriously depleted livestock in all districts.

The black majority government which came to power in 1980 introduced the Education Act of 1980 which emphasised free and compulsory primary and secondary education for all. According to Mamoyo (63) from Bulilima, this policy meant that pastoral children who had helped their parents in household and pastoral chores at home had to go to school leaving these chores mainly to their mothers. Mabhena (78) from Binga added that this period also saw a boom in industry as the economy opened up to the outside world in response to both independence and the lifting of sanctions by the international community. Most of the remaining males in pastoral communities drifted to towns for employment since both brain and brawn power were in demand. This caused sharp labour shortages in most pastoral homes and women had to carry the full load of pastoral responsibility.

The 1982 Legal Age of Majority Act was a milestone in opening up pastoralism to women as it allowed women to own land and other property in their own right. For the first time they could also register these in their own names. This act also gave women contractual rights which allowed them to engage in barter trade; for example, Madonga (68) from Matobo district said in 1987 she exchanged 5 goats for a cow and this was how she had managed to raise the 16 head of cattle she had in her name at the time. The act also liberated women to participate in wage labour either in Zimbabwe or in neighbouring South Africa or Botswana. This new-found liberty, however, benefitted mainly single and widowed women who do not need to seek husband's approval for their decisions. Movement of both males and single women increased after South Africa attained majority black rule in 1990.

The Fast-Track Land Reform Period (2000-2011)

The Fast-Track Land Reform Programme (FTLRP) in south-west Zimbabwe was not as euphoric as in other parts of the country. The reason could be because the government's thrust of this policy was more on land for crop production than for livestock rearing for which the region is best suited given its dry and hot conditions as well as good natural pastures. Mpofu (47) from Nkayi observed that the programme seemed to benefit war veterans' more than ordinary people in the community. MaDube (68) from Bulilima complained that the 5ha piece of land given to each household under the A1 model of the land reform programme was too small for cattle rearing, emphasising that cattle need large areas which allow them free movement. She said these small areas are too crowded and restrictive, reminiscent of the colonial period. Ndlovu (42) from Gwanda and Sibanda (43) from Beitbridge complained that cattle rustling increased as a result of the FTLRP adding that most of these newly settled areas are poverty stricken and this promotes criminal tendencies. MaNkala (46) from Bulilima complained that cattle herders whom they employ sometimes clandestinely sell cattle and report home that the cattle have got lost or stolen. The absence of cohesive traditional local power structures in newly resettled areas also promotes stock theft in these areas. This theft is also encouraged by the insatiable urban meat market in Bulawayo, Gwanda, Plumtree, Beitbridge and Hwange.

MaMoyo (61) from Bulilima, however, praised the FTLRP saying it availed farmers more land with virgin rangelands and better dip tanks. Magená (46) from Nkayi also appraised the programme saying it enabled farmers to cross-breed their indigenous *Tuli* and *Nguni* cattle with a variety of exotic breeds and this has resulted in better breeds. He added that newly resettled farmers are relatively young people who easily accept and adopt modern technology which improves both crop and animal yields.

The FTLRP further opened women's access and control over livestock. For example, women can now access input loan schemes from agricultural banks like Agribank. Under the FTLRP the government specifically and deliberately allocated 20 per cent (Zimbabwe National Gender Policy, 2004) of all land distributed in all districts to women. MaDube (65) from Matobo expressed gratitude to the government adding that she is managing to look after her AIDS-orphaned grandchildren from proceeds she gets from her new plot where she raises goats, cattle and also practices some crop farming. She cited the Deeds and Registration Act of 1985 as a progressive piece of legislation under which women are allowed to register property in their own names like she had done adding that it reduces women dependency on men.

Manifestation of Climate Change in South-western Zimbabwe

Respondents in all districts reported an increase in temperature for both summer and winter seasons over the years. Rainfall has generally decreased and rainfall patterns are both unpredictable and unreliable. For example, Gogo Ndebele (73) of Tsholotsho said the first rains (*imbolisamahlanga*) were normally received mid-September with the early rains proper (*insewula*) coming in mid-October. It was these early rains and not the first rains which were used for land preparation. The month of November was traditionally known to be a rainy month during which farmers planted their crops with a lot of hope. Although the rains would take a break in early December, the Christmas week normally received rain blessings. The month of January traditionally would not get much rain but would allow plants to grow. February was known to be the wettest month of the year according to MaBhebhe of Tsholotsho and these rains would continue until the end of the rainy season in March.

Gogo Mathwala (80) of Matobo observes that such a predictable weather pattern is no longer observable and farmers can hardly plan. This increases their vulnerability. She further described the nature of rainfall as violent and destructive since in most cases it is accompanied by very strong and violent winds, lightning and thunder or hail. Lightning and thunder casualties in the form of property, people and livestock are recorded every rainy season. Gatsheni (42) of Mwenezi said the agricultural season used to start mid-October but now serious farming starts in November. He observes that droughts are now more frequent, adding that they now experience more droughts than rain. Droughts occur every 3 - 5 years. Floods are generally not common in the region but when they occur, they are severe and extremely destructive to people, property and livestock, like the Cyclone Eline of 2000.

Magadlela (53) of Matobo observes that rain-bearing wind patterns are no longer predictable. In Matobo, rain-bearing winds usually blew from the south but currently this is no longer consistent as rain-bearing winds can blow from any direction.

Women, Climate Change and the Pastoral Economy in South-West Zimbabwe

According to Ravindranath and Sathaye (2002), reliance on pastoralism as an economic activity is in itself adaptation to climate change. Types of livestock observed in all the districts in the study area are cattle (*Tuli* and *Manjanja* breeds), goats (*Mie* breed), and donkeys with sheep found in fewer households. These breeds are adaptive to both climate change and variability. Sheep are fewer because according to MaNdlovu (64) from Gwanda, they are easy prey for dangerous animals and they do not make noise when being attacked to alert people of impending danger. Sheep also get lost easily since they can hardly re-track their way back if they pasture a distance from home. These reasons mean that sheep demand more labour than other livestock since they require close attendance almost all the time.

Respondents in all districts conceded that they have been pastoralists since early times and that they practise sedentary pastoralism. Pastoralism in southwest Zimbabwe is confined to Zimbabwean borders (it does not spill-over into Botswana or South Africa). However, it can be inter-district if there is a grazing agreement between chiefs in neighbouring districts. Pastoralism is the major economic activity in this region. Possession of livestock ensures sustained livelihood compared to any other economic activity. With livestock, one can pay school fees for the children, plough and cultivate the fields and get manure to improve soil fertility. Livestock can also be exchanged for grain and other goods and services.

MaNdlovu (62) from Gwanda summed up the importance of livestock when she conceded that "livestock are security against natural disasters like droughts, floods and pests". Madlela (60) also from Gwanda emphasised the importance of livestock saying: "It's better to have livestock than to do any other economic activity because through livestock one gets everything else, for example, I am currently extending my house through cattle sales". He added that in 2008, his family, like many others in his community, survived the hyper-inflationary period by exchanging cattle and goats for food. MaSibanda (70), a widow from Tsholotsho, also highlighted the economic and social significance of livestock thus: "*Ukufuya kuqakathekile kakulu. Ofuyileyo nguye umunthu osatshayo esigabeni*" (Livestock possession is what makes a home. One who owns livestock is respected and revered by the whole community).

It is evident from the confessions of the respondents, both men and women, that livestock are the economic base upon which communities in south-western Zimbabwe rely. The majority of the respondents, however, indicated that pastoralism is no longer pure but is now being supported by other economic activities. In Matobo District, for example, people compliment pastoralism with small-scale market gardening and *mopane* worms (*amacimbi*) harvesting which they sell in the neighbouring regional city, Bulawayo.

Focus group discussions from all districts revealed that they compliment pastoralism with subsistence growing of drought-tolerant crops, like sorghum and millet, which constitute their staple diet. In Bulilima and Matobo districts, *mopane* worms (*amacimbi*) harvesting is also an important economic activity which, besides satisfying household relish needs, is also practised commercially whereby the worms are processed by women and sold to Bulawayo and other areas throughout the country. The wetter parts of Matobo District, for example, Nathisa, Gulathi, Silozwi and Dobi which are within the culturally famous Matopos Hills, practise small-scale market gardening also for sale in Bulawayo.

Gold panning as a complimentary economic activity to pastoralism was reported in Gwanda District along Tuli and Shashe rivers and around Blanket, Freda and Vubachikwe gold mines. In Mwenezi and Beitbridge districts, gold panning was reported along Runde River mainly around Maranda communal areas while in Matobo District, it is common along Mzingwane River. Gold panning is common in these regions because Gwanda and Matobo districts are geologically gold regions while Mwenezi and Beitbridge have rivers whose sources are on Zimbabwe's mineral rich Great Dyke which itself is a rich gold belt. People in these districts practise mainly alluvial gold panning. This diversification is a strategy for adaptation to climate change and variability.

Respondents in all districts also mentioned in FGDs that hunting was an important supplementary economic activity to pastoralism but bemoaned the fact that wild animals were now very few and further apart mainly because settlements have encroached into their traditional hunting forests due to population increase. People around Matopos National Park, Chirisa National Park (Binga District) and private conservancies also complained that most wild animals have been confined to national parks, game reserves and conservancies and have, therefore, either become state property or private property depriving communities of their historical and traditional rights to their God-given common property by virtue of being endemic to these animal-rich areas. Communities around these protected areas practise illegal hunting (poaching) to satisfy their meat and other wild animal-related needs like skins, horns and other trophies. They mainly kill small wild animals like hares, springboks, impala which periodically stray from the park. The brave hunters, however, risk their lives by sneaking into the park and killing big game like kudus and antelopes. This has resulted in the incarceration of a number of such hunters with some being shot dead by game rangers. MaDube (63), a widow from Matobo District, observed that, "hunting is illegal but out of desperation people risk being arrested and even death". MaMoyo (40) a single mother also from Matobo District lamented saying that her father has been in and out of prison for hunting. Relations between pastoral communities and national parks rangers are thus very strained.

Respondents in all districts highlighted the importance of wage-based employment as another necessary complimentary activity to pastoralism, thus formal employment is encouraged because it enables them to buy more cattle, pay herders, afford dipping fees as well as buy other household necessities which do not warrant selling of livestock. While some people seek employment in the few industries still operational in Bulawayo, most of them, especially the youth, seek employment either in neighbouring South Africa or Botswana in what Matsa and Matsa (2010) referred to as the region's "Look-South Policy".

Women, Food Security and Adaptation

Issues of food security and management are an adaptation mechanism (Ravindranath and Sathaye 2002). A significant number of women (60 per cent) are *de facto* family heads for most of the year in south-west Zimbabwe since most able-bodied men either work in neighbouring Botswana or South Africa and come home for very short periods during the December Christmas holidays (about 2 weeks) and/or during the April Good Friday and Easter holidays (about 1 week) (Matsa and Matsa 2010). Other women are also *de jure* family heads either because they are divorced or widowed. This makes these pastoral women key schemers and executors of household level survival strategies in south-west Zimbabwe. According to MaDube (48), from Bulilima, children, who traditionally used to help women with these chores in the absence of their fathers, now attend schools and are away from home for most of the day (6 am to 6 pm) since schools are very few and far from home. Some children actually stay at schools where they have constructed rudimentary shelter (bush boarding schools) for themselves to counter distance. They only go home at weekends and/or during school holidays.

Indirect uses of cattle cited by respondents in FGDs include milk provision and meat when they are killed mainly during traditional ceremonies like weddings, funerals and appeasement of spirits, or when they die from drought-induced famine. Respondents revealed that cattle are a symbol of wealth, influence and power among pastoral communities. They are rarely killed for any culturally or traditionally plausible reason, even when they become visibly sick and feeble. The owners are always hopeful that they will recover.

Women play an important role in food processing and management. MaNcube (53) from Beitbridge cited milk as an important indirect product from cattle in pastoral communities. She explains that she uses fresh milk as relish, for feeding infants, for tea and for medicinal purposes, for example the treatment of stomach acids and to induce vomiting in order to clear ingested poison. It is also used for red eye treatment. By-products of milk include milk cream (*ulaza*), sour milk (*ihiqa*), thick sour milk (*amasi*), whey (*umlaza*) and milk body lotion (*umfuma*). MaNcube further clarified that although milking is mainly done by males, all milk products and milk by-products are the responsibility of women. In districts like Bulilima, Tsholotsho, Matobo, Mwenezi and Beitbridge, women are also active in milking. This is because these border districts have a high male emigration rate to Botswana and South Africa.

Milk processing is traditionally the responsibility of women according to MaNgwenya (72) from Binga. Observation revealed that women prepare calabashes, clay pots and other utensils used in milk processing. Soon after milking, milking calabashes (or cans) are cleaned by women. They then sieve the milk to remove dirt. If milk is to be served while still fresh, it is the females who boil and serve it. Women are also knowledgeable and skilful in further handling and storage of the rest of the unboiled milk for future use. They pour this milk in large fermentation containers (calabashes and clay pots).They then store these containers in areas/places with conducive temperatures which apparently only women can discern. These places can be in the kitchen, reserved rooms or outside for part of the day. Women also know when and how to remove cream and whey – stages which leave or form thick sour milk (*amasi*). Cream is usually removed after 2 - 3 days depending on temperature and season. Milk cream is used in place of cooking oil for preparing all types of vegetables, both fresh and dry. It can also be used to produce a traditional type of body lotion (*umfuma*) which is used by the whole family. *Umfuma* is produced by thoroughly boiling milk cream until the milk smell disappears and then allowing it to cool before use.

Like milk cream, whey is also removed from sour milk and what remains is thick sour milk (*amasi*). *Amasi* is used as relish for the staple *sadza* (traditional thick porridge prepared from maize, millet or sorghum meal). *Amasi* can also be served with cooked pumpkins, sweet potatoes or pounded and boiled finger millet (*umcaba*). In sour milk proper, whey is not removed but mixed with *amasi* to increase quantity. It is thus commonly served in large poor families where *amasi* may not be enough to satisfy all family members. According to Sebata (65) from Beitbridge, under normal circumstances, whey is fed to cats and dogs but in times of famine it becomes an important part of community diet.

Beef is another major direct product of cattle in pastoral communities. However, Mathwala (80) from Matobo and MaMoyo (63) from Bulilima explained that specific organs are traditionally allocated to different members of the family depending on their gender, age, and position of authority in the family. Organs are also allocated according to their traditional symbolic meanings and beliefs associated with them.

Men are normally allocated organs like *inhloko* (head), *amangqina* (leg), *isingqe* (behind), tail, liver, kidneys, *amalulu* (small intestines) and tongue. According to Mathwala (80) from Matobo, *inhloko* (head) is specifically served to males because they are family heads or will become family heads. *Amangqina* (legs) are also for males because they are believed to strengthen joints and to enhance male sexual activity. *Isingqe* (back) is believed to have soft quality meat hence is given to male family heads because they are special in the home. The liver is symbolic in that it denotes bravery which every family head must exhibit to be able to protect the family. Ox-tail and kidneys are also believed to be quality meat to be served to family heads as a sign of reverence. Mathwala further conceded that there is a strong belief that small intestines and the tongue should not be eaten by women and children because small intestines are believed to increase vulnerability to colds while the tongue should only be eaten by someone who can keep secrets hence both organs are served to men who are believed to be stronger both physically and morally. These beliefs conform to gender stereotypes which perceive women as weak vessels and gossipers. The reason why small intestines are reserved for male family heads may be because they are cleaner than the colon. Like the tongue, small intestines are also tender and tasty.

Organs specifically allocated to women include brisket (*ingiklane*) and *sadijana*. These parts are both fatty soft and tasty. They are a symbol of good living and hence, are believed to give satisfaction to women.

Boys are allocated the colon, the rectum and the pancreas. According to MaMoyo (63) from Bulilima, since boys take part in the slaughtering of cattle, these parts are normally roasted and consumed at the place of slaughter. They rarely get home, hence, their preparation is exclusively done by boys. The udder and male reproductive organs (testis and penis) are specifically reserved for boys and girls who have reached puberty and also to elderly males and females who have reached menopause. The traditional reason according to Mathwala (80) from Matobo is in order to control sexual immorality within the community. Any other part of the beast apart from those discussed above can be eaten by any member of the community.

Respondents from Tsholotsho, Bulilima and parts of Matobo revealed during FGDs that *inhloko* (the head) and *ububende* (blood) are prepared and served by both men and women while those from Nkayi and Gwanda conceded that these parts are exclusively prepared and served by males. *Ububende* (blood) is boiled cattle blood mixed with pieces of *ulusu*,(lungs) and parts of *isadijana*. This delicacy should be prepared within 2 days of slaughter to ensure that it does not rot.

Although slaughtering and skinning of the beast is done by men, all other meat preparations, for example, cutting of meat into strips for drying or into pieces for cooking is done by females. Storage of dried meat and its consumptive management (or rationing) is also done by women.

Important products from goats are meat for consumption and milk for tea. In Bulilima, and Tsholotsho, goat milk is also fomented into sour milk to be used as relish by the whole family. It was observed that milking of goats is done by women and children and this made goat milk to be easily accessible. Processing of the milk is also done by women. Goat milk needs thorough boiling to remove goat smell from it. Goat meat processing, that is, cutting it into strips for drying and its management is the responsibility of women. Specific goat meat parts, like that of cattle are allocated to members of the family for consumption.

Gender and Livestock Ownership in Pastoral South-West Zimbabwe

All respondents revealed that livestock ownership is still male dominated. Magadlela (53) of Matobo noted that it is not common for women to own cattle save for *inkomo yohlanga* (cow given to women as bride price for their daughters). Gogo Mabhena (78) of Binga pointed out that women ownership of livestock especially cattle is not feasible adding that women and whatever they own belong to the husband. Women can be given cattle by their parents as inheritance, for example, *inkomo yochago* or can buy their own cattle using income from grain selling, but as long as they are kept by men in their kraals, it is not known that they own cattle.

Research revealed that more single women own livestock than married women although their cattle are always fewer than those owned by men. Ways of accessing livestock, especially cattle, were open to single than to married women. Single women, especially those working in Botswana and South Africa, can buy cattle of their own. For example, Mabhebhe (42) of Tsholotsho pointed out that she had 4 cattle of her own. MaDube (41) of Matobo also revealed that after divorcing with her husband she went to work in South Africa and bought cattle of her own.

Research also revealed that more single women than married women registered cattle in their own names. Respondents from Tsholotsho and Bulilima pointed out during FGDs that some married women are beginning to register cattle in their own names, especially those with husbands working in South Africa. However, they were quick to point out that the registration does not symbolise ownership of cattle. This is done to make cattle clearance easier for women in times of cattle sales on the market so that they get money for school fees and other household requirements. Decisions concerning the selling of these livestock still remain with husbands.

All respondents revealed that generally, women own more goats than any other livestock. They also make decisions concerning the slaughter or sale of goats. This is because goats are easily accessed and goat ownership by women does not threaten patriarchal power. Ownership of cattle is a source of both economic and social power to males especially in rural areas. MaMoyo, (63), of Bulilima pointed out that women can exchange grain for goats.

Livestock Management by Women

Although Ravindranath and Sathaye (2002) argue that changing or diversification of gender roles is a strategy for adaptation to both pastoralism and climate change, respondents from Gwanda, parts of Matobo, Lupane and Nkayi felt that women should not be actively involved in livestock management like herding, dipping and treatment. They pointed out that women have engaged in these masculine roles because of the shortage of male labour. Accordingly, Magadlela (53) from Matobo and Ndlovu (50) from Gwanda, pointed out that it is unfair and cruel for women and girls to herd, treat and drive cattle to dip tanks. They argued that, if there is a shortage of male labour, herders should be employed. However, respondents from Beitbridge, Tsholotsho, Bulilima, Mwenezi and Binga revealed that roles are being exchanged. Women play a critical role in livestock management. The majority of women are available in rural areas compared to men. Children attend school – making it mandatory for women to manage livestock.

Gogo Mathwala (80) of Matobo maintains that she headed, milked and yoked during her childhood because in their family there were more girls than boys and boys were sent to mission schools. MaSibanda (61) of Tsholotsho concurred saying she also grew up herding cattle and even today her daughters-in-law herd cattle, too. "*Mina ngakhula ngiselusa lalamuhla omalukazana bami bayelusa*" (I grew up herding cattle and even today my daughters in-law also heard cattle). She further noted that today more women in their area herd cattle because children go to school and males are employed in towns, mines and some are in South Africa. Besides herding, women may employ herders and supervise them. Women may drive cattle long distances to dip tanks and in search of pastures and water. In drier regions where there are no dams and perennial rivers, women fetch borehole water for livestock and this is a laborious and tiresome job. Gogo Ndebele (73) of Tsholotsho pointed out that they dig trenches near boreholes or pour water into dishes for cattle to drink. This was supported by Mudimba (56) of Binga, who revealed that in their area women either get water from boreholes or dig wells along Sikalala River and pour it in trenches locally known as *mutiba* in the Tonga Language. MaDube (49) of Matobo notes that dams have silted, for example, Kafusi and Njadani along Shashane River, so livestock depend on borehole water. She boasted that women do everything that men can do.

All respondents revealed that livestock treatment was also heavy and dangerous work for women because it demands physical power. For example, the ox or cow has to be tied with a rope if sores are to be cleansed. They, however, argued that women either do it or supervised herders doing it. If they use modern medicine, treatment is mostly done by Agritex government officers, especially in newly resettled farming areas. In communal reserves, people acquire their vaccines and do it on their own. Dosing, inserting brand marks (*umthiso*), cleansing of sores and spray dipping is mainly done by males. However, traditional treatment for black leg and miscarriages (*ukupunza*) can be done by females. Mamoyo (61) of Bulilima noted that women in her area are like men (*Ngabafazi madoda*). They can do anything concerning livestock management. Magadlela (53) of Matobo pointed out that treatment for black leg (*ihlwili*) is done using a certain root commonly found on anthills. The root is chopped (*ukugxotshwa*) and mixed with a 20 litre bucketful of cold water to clean the intestines and to de-clot blood (*ukuchita igazi*).Cattle are made to drink the contents. MaDube (49) of Matobo conceded to having participated in giving cattle *ihlwili*. MaSibanda (61) of Tsholotsho revealed that the treatment for black leg besides *ihlwili* can be done using tree barks of *umvagazi* whose log is laid across the gate so that once cattle cross it; they cannot be affected by the black leg. In cases of miscarriages (*ukuphunza*), cows are made to drink *isihaqa* as treatment. According to Ravindranath and Sathaye (2002), pest outbreak may be a result of climate change so the development and use of indigenous livestock medicine is a strategy for climate change adaptation.

There are traditional specialists who are experienced in helping out livestock with complications in giving birth. MaSibanda (61) of Tsholotsho revealed that a log can be pushed inside the uterus to remove the placenta bit-by-bit while MaNdlovu (59) of Lupane pointed out that in order not to harm the beast, a hand is dipped into soap or into a local plant *(inkhunzane)* so that it becomes slippery. It is then pushed into the uterus to drag the placenta and the process is called *ukukhokhela*. Other respondents suggested that an indigenous plant *isikhukhu* is mixed with water and given to the cow to drink. Part of it is poured into the uterus, then the placenta is pulled out bit-by-bit. Both men and women can assist livestock in giving birth.

All respondents conceded that care of young ones and the sick lambs and calves is the responsibility of females than males. The orphans and the sick can be given porridge or assisted to feed (suckle) from other cows until they are able to feed on their own. This can be done by both males and females. In case of rains and colds, women keep these in kitchens with fire for warmth.

Women, Supplementary Feed and Water Management

During drought periods when there are no pastures women drive cattle (or they may instruct herdsboys) to far off places where pastures can be accessed (*emilageni*). The problem, however, is that such areas are rapidly shrinking because of population growth and the resettlement programme. Those who can afford buy salt which they add to traditionally prepared livestock fodder or they buy additional feed for their livestock. This is not always sustainable as stock feed is unaffordable for the majority of pastoral families. Many families preserve maize stalks and also dry groundnuts plant stems which they later use to feed their livestock in the dry season. When there is a need to serve some ailing livestock from possible death, Gogo MaSibanda (61) of Tsholotsho revealed that she would mix salt with mealie-meal and grain which she would use to feed the victim animal. Respondents from Beitbridge, Matobo and Bulilima reported that they give their livestock additional grass which they normally prepare during the rainy season when grass is usually plentiful. Livestock is also fed with wild plants and fruits like *umtshatshatsha, ihabahaba*, baobab, *inkelo* among others. Most households in all districts, in addition, preserve pumpkins, sweet reeds and *amakiliwane* for livestock feed during dire times. Although these measures are effective in reducing livestock mortalities, they are not always adequate as there are some years of serious drought when all these interventions are impossible since most of them are water-dependent to a very large extent. This was reported by MaNdlovu (51) from Binga.

Floods, although not very common in most districts, are a serious disaster whenever they occur, like in 2000 when the region was hard-hit by Cyclone Eline. Flooding of kraals is, however, a problem every normal rainy season. MaNtini (52) from Siphepha area in Tsholotsho explained that they as pastoralists counter this by shifting kraals to drier areas, or they may open trenches from the kraals to allow water to flow out, or they lay heaps of grass and tree branches in the kraal to reduce the effect of water on cattle. Women, as *de facto* family heads, organise work parties for all these remedial measures.

Constraints Encountered by Pastoral Women

Responses from FGDs showed that grazing lands are a problem; they are shrinking, especially in communal reserves. As a result women travel long distances in search of pastures. Respondents from Nkayi, Binga and Tsholotsho bemoaned the fact that this is dangerous for women as there are robbers, stock thieves and dangerous wild animals from game reserves. Those from Matobo and Bulilima highlighted that the areas are mountainous and hilly. In times of drought herders in these areas have to climb mountains and hills for pasture and browse and this is not safe for women.

Pastoral communities have to rely on employed herders who are unavailable, expensive (they are paid in the South African rand), and unreliable. Gatsheni, (42), of Mwenezi noted that herdsboys are hard to come by because they also want to go for better jobs in towns and to cross borders to Botswana and South Africa. Women pastoralists also have burdens added onto their household duties (reproductive roles) and additional labour of supervising herders. Hiring of labour, although expensive, was viewed by respondents as a way of adapting to labour demands. Gatsheni added that in the absence of herdsboys, women do the herding themselves.

Parts of Matobo and Gwanda had no problems with water. Water for people and livestock was reported to be available in Thuli, Shashe, Shashene, Mzingwane and Nyankuni which are perennial rivers. Mangwe Dam in Gwanda is another reliable source of water in Gwanda District. However, respondents from Bulilima, Nkayi, Beitbridge, Mwenezi, Tsholotsho and Binga highlighted water problems, Binga and Tsholotsho being the hardest hit. Both people and cattle in these districts rely on borehole water. The rivers are seasonal or unavailable and dams have silted. Boreholes are few and apart. Pastoral women travel long distances to the few silted dams, few boreholes and the drying rivers. Women also do the strenuous job of pumping out water from bush pumps (*ukukhotshosela inkomo amanzi*) into trenches and dishes. This demands physical power and increases women's burdens. Mudimba (46) of Binga observed that rivers like Sakalala are seasonal and there are no nearby dams. Observation showed that boreholes are few in all districts studied. Wells are dug (*ezihotsheni*) and women pour water into traditionally made trenches (*mutiba*) for livestock to drink.

Ncube (64) of Nkayi revealed that they travelled long distances in search of water in dams like Alicedale. MaBhebhe and Gogo Ndebele of Tsholotsho, reported that timetables have been made to rotate access to borehole water for both people and livestock. For example everyday of the week a certain number of households fetch water for household uses from 3am to 10am. From 10am onwards each day the same households fetch water for livestock. Three days a week are spent on fetching water for people and livestock. All this water fetching is done by women.

Problems encountered by women in relation to livestock treatment are premised on the fact that livestock treatment demands physical power. It also requires modern and traditional knowledge. Gwebu (47) of Gwanda noted that sometimes pastoralists have to invite a veterinary officer. However, medication is sometimes scarce and must be bought. Gatsheni (42) of Mwenezi conceded that treatment needs technical know-how which women may not have. Ndlovu (50) of Gwanda indicated that grandparents and men do not pass on to their children and wives knowledge relating to livestock management. This was corroborated by Gogo Mabhena (78) of Binga who noted that sometimes the family head (father) might have treated cattle using a certain traditional medicine either to increase the number of cattle or to protect them (*umthuso*) but they keep the knowledge a secret. This makes traditional medicine both unsustainable and unreliable.

In communal reserves, some dip tanks are far away or dilapidated. Women pastoralists have to travel long distances to functional ones. Livestock are also difficult to control. Sibanda (46) of Beitbridge confirmed that dipping infrastructure is either dilapidated or unavailable. Dipping chemicals are expensive. Pastoralists are made to pay a dipping fee ranging from 20 US cents to US$1 (in Bulilima District).

The majority of women pastoralists are unemployed. Madlela (84) of Gwanda and Ngwenya (74) of Matobo felt that it's not secure for a woman to drive cattle to dip tanks. Women can be targeted by immoral men and harmed.

Like dipping, kraaling was viewed as a strenuous job for women although women do it in pastoral communities. Calves have to be separated from their mothers for milk purposes. Logs (*imiqoqo*) or branches (*amahlahla*) are used to close kraal entrances. These logs are not only heavy but also dangerous for women.

In all the districts, this was reported as becoming uncommon. The reasons given were that nearby grazing lands are no longer available and there is labour shortage. The period before 10am, women are burdened by reproductive roles of household chores (fetching water, firewood and cooking) and child rearing (bathing, feeding and washing for infants). Gwebi (47) of Gwanda noted that early morning grazing (*ukuchakisa*) has been compromised because women have to manage household work and also do men's duties – leaving no time for it. The result has been low milk yields in the home.

The outward drifts of men to towns, South Africa and Botswana have caused severe labour shortages in all districts. This has increased women labour burdens.

Men have traditionally been secretive about their indigenous knowledge. Their absence from the home has meant a shortage of skills in livestock treatment and management.

Most respondents, however, revealed that men's absence may be advantageous in that employment in towns and neighbouring countries enabled pastoralists to buy more cattle and to save the ones they already have since they will not have to sell them to cover for pressing family needs like clothes, fees for the children and other such requirements.

Conclusion

Communities in south-west Zimbabwe have been pastoralists since early times and pastoralism is still a highly valued economic activity today. Their traditional livestock are cattle, goats and donkeys – all of which are still valued. However, pastoralism is now complimented by other economic activities like cultivation of drought-tolerant crops, harvesting of *mopane* worms, wage employment and gold panning in some areas. This is in line with Ravindranath and Sathaye's (2002) view that diversification of livestock, crops and economic activities are a mechanism for adaptation to climate change and variability.

Livestock are significant for both direct and indirect use. Direct uses range from draught power, barter trade, cash and traditional ceremonies while indirect products, especially milk and meat are crucial for nutrition and other household uses. However, the preparation, processing and consumption of milk and meat are gendered. Each gender is assigned and consumes specific parts of the beast. There are both positive and negative social constructions relating to certain parts of the beast, gender and age. Thus, food preparation and management is important in minimising adverse impact on food security and livelihoods (Masika 2002).

During the pre-colonial period, pastoralists had more heads owing to favourable conditions relating to availability of water and vast rangelands (Hargreaves, Bruce and Beffa 2004). However, pastoralism was male-dominated. Participation of women in livestock management was minimal.

During the colonial period, livestock herds dwindled as a result of colonial polices of de-stocking African herds and expropriation of land that saw pastoralists being crowded in unproductive and poor grazing lands. However, women became active and full time pastoralists owing to labour shortage resulting from wage employment and schooling. Employing herders is an adaptation strategy to ameliorate labour shortages as postulated by Ravindranath and Sathaye (2002).

The post-independence period has witnessed an increase in women participation not only in livestock management but also in ownership and control. Women have started to own land and livestock in their own right. Women have also begun to access input loan schemes and to register property including livestock in their own names.

Pastoral women have benefitted from the Fast-Track Land Reform Programme which allocated 20 per cent of land to women (ZNGP 2004).

South-west Zimbabwe is experiencing climate change whose manifestations are uncharacteristically high temperatures, unpredictable rains, increased droughts and floods as well as inconsistent wind patterns. This is consistent with climate change studies' results from southern Africa and Zimbabwe by Hirji, Maro and Chiuta (2002) and Zhakata (2008).

However, pastoralists whose majority are females have come up with adaptation and coping strategies in the face of climate change and variability to meet their livelihood needs. These include diversification of economic activities and taking over roles traditionally conceived as masculine.

Although in the post-independence period women own livestock, it must be noted that patriarchy is still dominant in pastoral communities. More single than married women own and register land and livestock especially cattle in their own names. However, women play a key role in livestock management: they herd, search for pastures, dip and treat livestock although these are masculine roles which are strenuous for women. Masika (2002) observes that the search for water, pastures and pen-feeding are all strategies of adaptation to climate change and variability.

Women face a number of constraints in livestock management which include shrinking pastures, shortage of water, shortage of labour, droughts resulting in scarcity of water and poor grazing lands. Sometimes their lives are at risk from robbers, dangerous animals, rapists and cattle rustlers.

Women stretch their labour to cover both productive and reproductive roles. They tend to lack modern livestock management know-how and mostly rely on traditional knowledge. Their adaptation and coping strategies to climate change are mainly spontaneous (unplanned) adjustments for survival and there is heavy reliance on locally available resources.

In times of droughts, livestock are driven to open rangelands or to salty anthills (*izimunyu*). Remains from the fields are preserved for the dry season, for example, grain stalks and grass are mixed with salt and given to cattle. Livestock can also be fed with tree fruits, tree bark, leaves and water melons to make them survive harsh dry periods and droughts. However, these coping strategies are not adequate to help increase livestock numbers to the levels they had in pre-colonial times. Major reasons given for low numbers of livestock were inadequate pastures, drought, shortage of labour and the commercialisation of the bride price.

Recommendations

Pastoralism in south-west Zimbabwe is a valued economic activity for livelihood and women play a key role in livestock management. For this reason, there is need for Zimbabwe to implement the 2010 Addis Ababa policies to promote the pastoral economy in the south-west of the country.

Land and livestock ownership rights should be democratised by including them in the country's national constitution and following up on their implementation mainly by gender sensitive stakeholders both in government and civil society. If these are not solved, women's effectiveness in pastoral economy will remain invisible.

There is also a need to conduct awareness campaigns and training workshops at community level in order to educate women on climate change and environmental management to reduce risks of vulnerability to droughts.

Information relating to livestock management should also be disseminated to pastoral women. There is need to facilitate extension studies for women, for example, education on animal health and use of technology relating to livestock management. Women should also be educated on water conservation measures as pointed out by Mogotsi, Nyangito and Nyariki (2011).

Both women and men at local level should be incorporated in decision-making, climate change, mitigation and adaptation initiatives and debates. This should lead to a shift from public adaptation initiated and implemented by governments to private adaptation which is initiated by individuals, households and communities themselves (Ravindranath and Sathaye 2002).

Pastoral communities also need to be sensitive on the value of preserving pastoral economic systems. The merging of IKS and modern knowledge is crucial to promote increased resilience of poor people's livelihoods, in this case, pastoral women and their families. This idea is shared by Masika (2002).

Physical infrastructure need to be developed for easy access by pastoral communities, for example, dams, boreholes, dip tanks and kraals, among others.

Above all, there should be an integration of gender analysis, research and gender awareness to motivate both men and women to address inequalities relating to issues of climate change and pastoralism (Cannon 2002).

References

Barret, J. C. 1992. The economic role of cattle in communal farming systems of Zimbabwe. Pastoral development network. In Hargreaves, Bruce, and Beffa. 2004. *Disaster mitigation options for livestock production in communal farming systems in Zimbabwe*. Bulawayo: ICRISAT.

Bhila,H. and Shilington, K. 1986. *A junior certificate history for Zimbabwe* Book 1. Harare: Longman.

Bowen, A. and Pallister, J. 2000. *AS level geography*. Oxford: Heinemann Educational Books.

Boyer, K. 2004. *Feminist Geography in the archive: Practice and method in geography and gender reconsidered*. New York: Rensselaer Polytechnic Institute.

Cannon,T. 2002. Gender and climate change hazards. In *Gender and Development*, vol.10, no.2, 45-51. Available online: http://www.adpc.net. (accessed on 31st October, 2011).

Carr, M. 1997. *New patterns: Process and change in human geography*. Cheltenham: Thomas Nelson and Sons Ltd.

Chandler, D.J. and Wane, N. 2002. Indigenous gendered spaces: An examination of Kenya. In *JENDA*: Online journal, Vol. 2, no.1,1-30.

Chenje, M. (ed.). 2000. *State of the environment. Zambezi Basin*. Harare: SADC/IUCN/ZRA/SARDC.

Chenje, M., Sola, L. and Paleczny, D. (eds.). 1998. *State of Zimbabwe's environment*. Harare: Government of Zimbabwe.

Conroy. 2002. *Socio-economic methodologies for natural resources research best practice guidelines*. London: Natural Resource Institute of Greenwich.

Davis, N.E. 1978. *History of Southern Africa*. Harlow: Longman.

Department of Rural Economy and Agriculture (DREA). 2010. *Policy framework for pastoralism in Africa: Securing, protecting and improving the lives, livelihoods and rights of pastoral communities*. Addis Ababa: DREA.

Ellis, J.E. 1995. Climate variability and complex ecosystem dynamics: implications for pastoral development. In Scoones, I. (ed.), 1998, *Living with uncertainty, new directions in pastoral development in Africa*. London: Intermediate Technology Publications.

Food and Agriculture Organization of the United Nations (FAO). 1995. *World agriculture towards 2010*. Rome: UN-FAO.

2004. Drought impact mitigation and prevention in the Limpopo River Basin: A situational analysis. *Land and Water Discussion Paper 4*.

Ford, M.J. 1983. The changing climate: Responses of the natural fauna and flora. *The Journal of Ecology* Vol.7, no.3,1027-1028.

Government of Zimbabwe. 2004. Zimbabwe National Gender Policy. Harare: Government Printers.

Hargreaves, S.K., Bruce, D. and Beffa, M.L. 2004. *Disaster mitigation options for livestock production in communal farming systems in Zimbabwe.* Bulawayo: ICRISAT.

Hirji, Johnson, P. Maro, and Chiuta, T. M. (eds.). 2002. *Defining and mainstreaming environmental sustainability in water resources management in Southern Africa*. Harare: SADC/IUCN/SARDC.

International Crops Research Institute for the Semi-Arid Tropics (ICRISAT). 2008. *Building adaptive capacity to cope with increasing vulnerability to climate change.* Andhra Pradesh: ICRISAT.

International Panel on Climate Change (IPCC) .2006. Guidelines for national greenhouse gas inventories: Agricultural forestry and other uses. Available online: http//www.ipcc.ch/pdf. (accessed on 25th October, 2011).

Kelly, P.M. and Adger, W.N. 200. Theory and practice in assessing vulnerability to climate change and facilitating adaptation. *Climatic Change*, 47, 4, 325-352.

Kraub, J. 2011. *Gender and climate change: Gender experiences from climate related GIZ Projects*. Bonn and Eschborn: GIZ.

Masendeke, D. 2008. Farmers *perception of climate change in Zimbabwe in ICRISAT building adaptive capacity to cope with increasing vulnerability due to climate change.* Andhra Pradesh: ICRISAT.

Masika, R. 2002. *Gender, development and climate change*. Oxford: Oxfam.

Matsa, M. and Matsa, W. 2010. Gender, resource management and the rural landscape in Africa. *Journal of Sustainable Development in Africa,* vol. 12, no.4, 153-163.

Merchant, C. 1995. *Earth care: Women and the environment.* New York: Routledge.

Mogotsi, K., Nyangito M.M. and Nyariki D.M. 2011. The perfect drought? Constraints limiting Kalahari agro-pastoral communities from coping and adapting. *African Journal of Environmental Science and Technology,* Vol. 5, no.3, 168-177.

Momsen, J.M. 2004. *Gender and development*. London: Routledge.

Moser, C. 1993. *Gender planning and development: Theory, practice, and training*. London: Routledge.

Nengomasha, Z; Homann, S., van Rooyen A. and Moyo T . 2007. Strategies to improve goat production and marketing: BASELINE Information for semi-arid Zimbabwe. International Crops Research Institute for the Semi-Arid Tropics. Bulawayo: ICRISAT.

Njiro, E.I. 1999. *Women's empowerment and the anthropology of participatory development. The feminisation of development process in Africa*. London: Routledge.

Olmos, S. 2001. Vulnerability and adaptation to climate change: Concepts, issues, assessment methods. Available online:http://www.cckn.net. (accessed on 29th May, 2012).

Prew, M. Pape, J. Mutwira, R. and Barnes, T. 1993. *People making history*. Harare: ZPH.

Ravindranath, N. H. and Sathaye, J.A. 2002. *Climate change and developing countries*. London: Kluwer Academic Publishers.

Rohrbach, D.D. 1999. Constraints and opportunities for women farmers in Southern Zimbabwe: Will women farmers invest in improving their soil fertility management. In Snapp, S.S. (ed.), Improving soil management options in Malawi and Zimbabwe. *Proceedings of Regional Workshop on Launching of a DFID Support Project*. Bulawayo: ICRISAT.

Rohrbach, D.D, van Rooyen, A.F and Hargreaves, S.K. 2004. *Disaster mitigation options for livestock production in communal farming systems in Zimbabwe*. Bulawayo:FAO and ICRISAT.

Schaefer, R.T. 2006. *Sociology: A brief introduction*. Boston: McGraw Hill.

Settel, J. 1999. *Exploding ants: Amazing facts about how animals adapt*. New York: Atheneum Books.

Shiva, vandana. 2011. *Interactive dialogue on harmony with nature*. United Nations General Assembly speech, Wednesday, 20 April 2011· NLB Conference Room 2.

Smit, B., Burton, I., Klein, R., and Wandel, J. 2000. An anatomy of adaptation to climate change and variability. *Climatic Change* 45, 223–251.

Tevera, D.S. and Moyo, S. (eds.). 2000. *Environmental security in Southern Africa*. Harare: SAPES.

United Nations Development Programme (UNDP). 2009. Gender and climate change: Impact and adaptation. IRADe. Available online: http://www.csd-i.org/. (accessed on 30th October, 2011).

USAID. 2007. *Adaptation to climate change. A guidance manual for development planning*. New York: USAID.

Van Rooyen, A. and Homann, S. (eds.). 2009. Enhancing incomes and livelihoods through improved farmers practices on goat production and marketing. *Proceedings of a workshop organised by The Goat Forum*. Bulawayo: ICRISAT.

Woolmer, W., Sithole,B. and Makamuri, B. 2002. Crops, livestock and livelihoods in Zimbabwe. In Hargreaves, S.K., Bruce, D. and Beffa, M.L. 2004. *Disaster mitigation options for livestock production in communal farming systems in Zimbabwe*. Bulawayo: ICRISAT.

Zhakata, W. 2008. Climate *change in Zimbabwe in ICRISAT building adaptive capacity to cope with increasing vulnerability due to climate change*. Andhra Pradesh: ICRISAT.

Index

abnormal community mobility 266
abnormal mobility 278, 291
Absuma 102, 103
adaptation 2, 4, 5, 6, 7, 8, 10, 12, 20, 21, 46, 53, 56, 66, 72, 76, 83, 89, 90, 93, 108, 130, 138, 155, 158, 161, 173, 190, 195, 200, 221, 225, 227, 228, 232, 233, 238, 246, 253, 259, 266, 269, 277, 285, 292, 294, 300, 306, 313, 325, 331, 338
adaptive capacity 4, 11, 13, 40, 76, 83, 90, 108, 127, 129, 191, 224, 229, 256, 273, 283, 300, 341, 342
Adaptive capacity 130, 271, 283
Adda 11, 83, 101, 102, 104, 109, 113, 114
Afar culture 11, 83, 109, 111
Afar ethnic groups 275
Afar men 11, 83, 114
Afar pastoral communities 105
Afar pastoralists 86, 87, 90, 91, 95, 105, 109, 111, 118
Afar tradition 102
Afar women 83, 86, 100, 102, 108, 112
African Purchase Areas 304
agricultural
  communities 190, 225, 226, 250
  cultivation 235
  extension services 307
agro-ecology 39
agro-forestry 172
agro-pastoral 12, 172, 187, 190, 191, 206, 259, 269, 278, 289, 301
agro-pastoral communities 12, 187, 190, 191, 206, 259, 266, 278
agro-pastoralism 39, 89, 179, 192, 208, 234, 278, 303
agro-pastoral women 191
agro-processing 239
alternative sources of income 53, 68, 128, 284
ancestral communal lands 304
ancestral spirits 201
animal husbandry 92, 99, 225, 301
Aqdi 102
Arid and Semi-Arid Lands (ASALs) 64, 68
Asia Tsunami 6
asset inheritance 102
autonomous adaptation 233, 237, 246, 253
Basic Education Assistance Module (BEAM) 191, 217
behavioural adaptation 309
Behaviour Changes 3
berkads 289
bio-diversity conservation 225, 238
bio-diversity resources 51
bio-physical changes 96
blank slates 198, 214
Borana pastoral group 89
brick-laying 239, 249, 250
Bush-burning 38
bush encroachments 266, 282, 283
bush fires 239
capacity-building 40, 76
care-giving tasks 194
casual labour 161, 234, 252
Cattle Corridor (ug.) 230, 238
Cattle rustling 315
cattle
  production 23, 306
  ranching 310
  tax 304, 317
  tillage system 304
Charcoal burning/production 35, 61, 73, 74, 76, 171, 174, 245, 250, 251, 281, 286, 291
childcare 207
child-headed households 209
child labour 217, 286
child mortality 177
children engagements
see esirata 74
Chivi District 12, 187, 191, 200, 203, 205, 208, 209, 212, 214, 215, 216, 217
civil society organisations 225, 237, 240
clan conflicts 69, 100, 267
Clean Development Mechanism (CDM) 4
Climate Change 1, 3, 5, 11, 17, 24, 29, 84, 87, 89, 94, 98, 100, 102, 105, 121, 127, 144, 155, 157, 161, 170, 187, 200, 203, 205, 223, 227, 229, 241, 265, 278, 282, 288, 297, 300, 306, 321, 322

Climate induced Shocks 85, 194, 268
climate
  adaptations 44, 119, 265, 269
  change-related diseases 139
  change-related information 112
  change-related problems 6, 20, 24, 28, 32, 33, 34, 42
  -induced changes 148
  related disasters 162
  -related hazards 266
  -sensitive economic activities 22
  -sensitive resources 5, 70
  shocks 5
  variability 1, 3, 6, 18, 20, 34, 39, 50, 88, 105, 122, 127, 132, 146, 151, 161, 188, 195, 200, 202, 205, 213, 216, 219, 227, 228, 268, 270, 271, 283, 294, 295, 299, 300, 312
climatic hazards 49, 50, 283
climatic stresses 35
collective-based survival strategies 234
commercial agriculture 298, 299, 305
commercial sex work 209, 214
communal
  farming 192, 340, 341, 342
  grazing 304
  rangeland resources 49
  tillage 306
Communal Reserves (zim) 304
community
  adaptation 194, 221, 285, 303
  based self-help 210
  gardens 208
  level projects 285
  migration 266
  owned natural resources 41
  policing 40
conflict management 38, 40, 42, 43, 80, 119, 265
conflict-prone areas 41
conflict-related problems 34
coping mechanism 35, 162, 209
coping strategies 7, 11, 17, 20, 21, 28, 47, 53, 66, 77, 112, 159, 161, 171, 174, 179, 190, 210, 234, 265, 274, 280, 306, 307, 337
crisis Zimbabwe 215
Crop cultivation 18, 32, 36, 42
crop diversification 6, 195
crop failure 42, 207, 266, 305
crop management practices 271
Cross-Border Jumping 212
cross-border trade 156, 210, 214, 216
cultural
  burdens 24
  embeddedness 198
  paradigms 198
  responsibility 52, 73
  status 51
Cushitic-speaking people 91
customary
  institutions 270, 273, 278, 290, 291, 292
  land-use agreement 278
  resource management systems 278
  systems 278
Cyclone Eline 202, 322, 333
cyclones 21, 190, 195, 205
decision-making politics 85
decision-making process 272, 284
deforestation 1, 18, 61, 62, 72, 96, 201, 209, 224, 238, 243, 251, 273, 288
depletion of resources 267, 283
desertification 2, 35, 188, 288
de-stocking 41, 105, 336
destocking 307, 317
Dhowrto/Dawarsi 288
disaster
  impact mitigation 40
  management 5, 196, 219
  risk reduction 38, 40, 42, 118, 119, 271, 272
  risks 5, 21, 35
diversification of livestock 234, 336
division of labour 12, 51, 85, 89, 103, 155, 157, 163, 174, 179, 180, 206, 207, 217, 229, 232, 243, 244, 248, 264, 279, 287
domestic burdens 265
drought
  induced shocks 85
  related shocks 108
  resistant indigenous crops 286
droughts 2, 7, 20, 22, 29, 37, 39, 40, 60, 87, 89, 101, 157, 160, 168, 169, 170, 177, 179, 190, 198, 202, 205, 206, 213, 230, 241, 276, 280, 281, 289, 291, 301, 322, 338
dryland bio-mass 273

dry land ecosystems 5
early marriage
see under-age marriage 103, 115
early marriages 209, 215
early warning systems 272, 284, 285
economic capital systems 187
ecosystem adaptation 238
ellas 29
El Niño 19, 168, 299
emergency aid 38, 41
environmental
challenges 198, 275
conflicts 274, 294
degradation 208, 224, 225, 245, 246, 251, 255, 257, 270, 273, 278
management 47, 196, 310, 338
parameters 161
sustainability 226, 341
Erer Gota River 267
esirata 64
extension agents 272
extra-pastoral economic activities 89
Fast Track Land Reform Programme 202
feeding systems 74
female-headed households 128, 231, 244, 287, 313
female malnutrition 177
feminisation of poverty 191, 217
Feminist geography 310
Fentalle Woreda 11, 121, 123, 131, 132, 134, 142, 146
financial capital 177
Fishing 243, 252
flash floods 31, 42
flood hazards 39
floods 2, 6, 12, 21, 30, 31, 32, 42, 157, 160, 162, 169, 170, 171, 173, 179, 188, 190, 202, 230, 290, 298, 301, 302, 308, 314, 323, 337
Focus Group Discussions (FGDs) 28, 30, 39, 55, 61, 76, 110, 164, 165
food-for-work arrangements 39
Food-for-Work Programmes 210
food
insecurity 22, 217, 236, 266, 269, 270, 286
procurement 122, 301
security 2, 15, 44, 48, 80, 127, 158, 182, 184, 195, 241, 247, 269, 274, 286, 301, 303, 325, 336
fossil fuel 297
fragile ecosystem 187, 192
gender
adaptation 306
barriers 160
based conflict 230
based differences 7, 85
based division of labour 103, 163, 179, 206, 207, 217
based inequalities 85
based violence 216
equality 44, 91, 167, 178, 179, 185, 220, 226, 295
inequalities 122, 179, 194, 310
neutral 225, 228, 306
neutral language 228
sensitive strategies 271
Gender-differentiated impacts 8
Gender-related risks 8
geo-political system 189
Global climate change 46, 156
global warming 18, 21, 297
gold panning 209, 324, 336
grain loan scheme 191, 217
grandmother-headed households 191, 209
greenhouse gases 1, 13, 18, 22, 114, 171, 224, 266
greenhouse gas-intensive technologies 3, 4
group labour 210
Gurgura clan 267, 292
health-related risks 18, 41
Herd accumulation 273
Herd diversification 34
herding-related activities 95
Herd management practices 273
Herd-sharing 50
herd splitting 89, 112, 234
high morbidity 246
HIV/AIDS 22, 71, 216
home-based care 206, 207
household
coping strategies 285
division of labour 103, 155, 157, 163, 174, 179
labour 89, 103, 105

Human agency 198
human
activities 1, 18, 122, 202, 224, 228, 243, 273
activity 127, 188, 227
demography 19, 31
induced climate change 270
land-use 188
security 181, 222, 266, 285, 293
settlements 274
humanitarian crises 271
hunting 303, 324
hut tax 317, 319
Ihefu Valley 157
Ihefu wetland 164, 172
illiteracy 167
income diversification 89, 109
income-generating activities 12, 37, 67, 70, 74, 76, 175, 177, 180, 223, 224, 237, 247, 253, 257, 280
income-generating groups 279, 291
indigenous customary institutions 292
indigenous knowledge 115, 219, 238, 254, 255, 261, 269, 284, 285, 308, 309, 335
indigenous survival mechanisms 267
individual wealth accumulation 11, 83
Institutional responses 18
inter-clan conflicts 100
inter-cropping 208
inter-ethnic-based conflicts 274
Inter-governmental Panel on Climate Change (IPCC) 1, 203, 227, 300
irrigation canals 267, 274, 275
Kajiado County 11, 45, 54, 55
kebele 123, 140, 143, 146, 149
Keyd 288
khat 280, 282
Kilosa District 12, 155, 156, 157, 164, 168, 178
Kyoto Protocol 3, 4, 13
labour
burden 229, 245, 253
intensive 237, 253
migration 11, 83, 90, 108, 113, 312
productivity 246
shortages 319, 335, 336
wastage 42
Land Husbandry Act of 1951 304, 317
Land
adjudication 49
conflicts 278, 304
degradation 62, 63, 171, 179, 239, 266, 286, 304
demarcation 61, 70, 74, 75
grabbing 278
management 112, 271
landslides 46
land tenure systems 150, 292, 304
livelihood
diversification mechanisms 269
enhancement 38, 42
shocks 276
livestock
density 31
foraging 49, 62
mortalities 332
loan schemes 307, 321, 336
locust infestation 266
Maasai women pastoralists 11, 46, 47, 48, 53, 62, 64, 66, 68, 73, 74, 77
magetta 275
masculinity crisis 215
massive migration 29
Matabeleland 298, 312, 313
maternal health 49
media famines 276
medicinal herbs 246
meticulous classification 94
micro-credit services 41
Millennium Development Goals (MDGs) 48, 309
mitigation 2, 4, 6, 40, 85, 91, 114, 148, 179, 196, 218, 240, 266, 271, 300, 301, 307, 338
mitigation initiatives 266
modern livelihood systems 49
modern livestock management 337
moral decadence 215
mud-streams 188
multiple cropping 208
Nakasongola District 226, 238, 244, 252, 253, 256, 260, 261
National Agricultural Advisory 251, 254
natural bio-diversity 192
natural disasters 6, 7, 18, 41, 52, 64, 122, 159, 160, 190, 192, 194, 196, 219, 265, 268, 301, 306, 309, 323

natural hazards 50
Ndebele pastoral communities 317
Nolle clan 267
nomadic systems 45
non-food items 30, 39
non-state actors 38
nutritional garden 205
Oldonyo-orok sub-location 11, 47, 54, 55
Oromia Regional State 25, 274
over-grazing 37, 96, 99
over-stocking 23
Participatory Rural Appraisal (PRA) 17, 28, 193
pastoral
  communities 5, 6, 8, 10, 12, 17, 18, 20, 21, 23, 24, 28, 29, 32, 34, 35, 36, 37, 40, 42, 75, 87, 91, 105, 121, 122, 143, 146, 148, 156, 162, 180, 187, 190, 191, 206, 223, 224, 226, 229, 230, 231, 232, 234, 235, 237, 238, 242, 244, 253, 257, 259, 261, 266, 267, 278, 289, 302, 304, 307, 308, 313, 316, 317, 318, 319, 324, 325, 326, 327, 335, 337
  crisis 189
  exchange networks 161
  families 159, 280, 291, 314, 332
  pastoral livelihoods 19, 23, 28, 84, 159, 275, 280, 290, 291
  mobility 34, 280, 291
  production system 23, 31, 273, 291
  related activities 225
Pastoral livelihood 98
Pastoral nomadism 303
Pastoral supportive activities 89
pastoralism-related policies 238
pastoralist women 7, 10, 12, 13, 88, 123, 135, 136, 138, 140, 143, 146, 147, 162, 165, 190, 191, 276, 277, 288, 312
patriarchal attitudes 309
peace-building activities 41
petty trading 11, 83, 89, 90, 252, 279, 286, 291
planned adaptation 90, 227, 233, 237, 246, 253, 307
planned adaptations 224, 229, 233
poaching 324
population growth 23, 46, 62, 74, 203, 270, 274, 332
post-disaster diminishment 268
poverty eradication 48
poverty reduction 38, 40, 77, 113, 167, 178, 180, 183
power indicators 90
power inequalities 85
property rights 283, 305
public education 40
publicl-owned resources 70
Rai 288
raiding of cattle 230
rain
  -fed agriculture 92, 170, 190, 270, 300
  -fed livestock-keeping 50
  making ceremonies 201
recurrent drought 88, 89, 92, 93, 99, 101, 105, 109, 168, 266, 275, 280, 282, 287, 291
redistribution systems 278, 288, 290
resettlement programme 332
resource-based conflicts 100, 111
resource management strategies 6, 34, 37, 194
resource ownership 157, 179, 180
Resource-Poor Households 288
restocking programmes (Ug.) 232
risky environments 50
river-based resource 267
rotational grazing 6, 195, 278
sanitation systems 285
School drop-out 209
Seasonal changes 59
seasonal
  climatic change 265
  migrants 128
  mobility 34
  ponds 239
self-aware actors 187, 191, 209
semi-arid regions 5
semi-arid tropics 298, 299
semi-humid tropics 303
semi-nomadism 63
sensitisation programmes 256
Services (NAADS) 251, 254
settled agriculture 276
slaughter of animals strategy 247

social
 capital 160, 266, 283
 dynamics 270, 277
 institutions 279, 288
 networks 50, 234
 obligations 50
 support mechanisms 34, 287
 welfare indicators 49
socio-cultural 'discrimination' 19
socio-cultural fabric 40
socio-cultural practices 18
socio-economic group(s) 17, 41
soil degradation 72, 76, 269
soil fertility management 271, 342
Somali customary institutions 290
Somali pastoral communities 285
South Omo 11, 17, 18, 21, 25, 30, 31, 32, 33, 35, 36, 37, 38, 41, 42
spatial mobility 265
spatial movements 135
spiritual appeasement 305
spiritual factors 216
Stall-fed animals 237
sub-Saharan belt 45
survival coping mechanisms 268
survival strategies 72, 73, 187, 208, 209, 210, 214, 216, 325
temporary migration
 See Seasonal Migration 223, 247
Traditional Astronomy 107
Traditional Livestock Production 13, 62, 77, 297
traditional
 beliefs 284
 cattle herders 52
 communal land 62
 coping strategies 7, 306
 traditional foods 74
 gender-based division of labour 205, 206, 207
 herbal medicine 176
 knowledge 81, 122, 163, 185, 284, 289, 299, 307, 334, 337
 livelihood 73
 pastoralism 303, 305
 pastoral management systems 305
 rites 201
 social security 217, 288
 social security systems 288
 values 96
 workload 156, 159, 280
Tree Planting 255
unconventional famine food 103
Under-age marriage 103
urban migration 171, 173, 209
valley dams 224, 239, 253
valley tanks 239
viscous cycle of poverty 170
vulnerability index 123, 124, 126, 129, 143, 144, 146, 147
wage labour 112, 236, 281, 286, 292, 320
water
 conservation measures 271, 338
 management 76, 255, 289, 307
 rationing 39
 scarcity 11, 33, 46, 47, 48, 49, 51, 53, 54, 56, 57, 59, 60, 61, 62, 63, 66, 67, 68, 69, 71, 72, 73, 74, 75, 76, 77, 156, 283, 288, 290
 scarcity-impacted irrigation 283
 shortages 2, 28, 52, 62, 202, 266, 306
 -stressed areas 158
watersheds 246
weather-related disasters 5
wetlands encroachment 251
wild animal-related needs 324
wildlife tourism 225
women empowerment 18, 41, 42
woreda 28, 40, 123, 124, 131, 133, 134, 136, 139, 141, 142, 143, 144, 145, 146, 147, 148
Zakat 287
Zambezi Basin 298, 299, 340
Zunde RaMambo (chief's granary) 217

www.ingramcontent.com/pod-product-compliance
Ingram Content Group UK Ltd.
Pitfield, Milton Keynes, MK11 3LW, UK
UKHW041144300926
15471UKWH00012B/127

9 789970 252367